THE
ELECTRONIC STEALING
OF THE
2004
U.S. PRESIDENTIAL ELECTION

ISBN: 1-4196-8242-3
ISBN-13: 9781419682421

Visit www.booksurge.com to order additional copies.

THE
ELECTRONIC STEALING
OF THE
2004
U.S. PRESIDENTIAL ELECTION

Democracy Hijacked Again

GARVIN KARUNARATNE
Ph.D. (Michigan State University)

OTHER WORKS BY DR. GARVIN KARUNARATNE

The Administrative Bungling that Hijacked the 2000 US Presidential Election, The University Press of America, 2004.

How the IMF Ruined Sri Lanka & Alternate Programs of Success, Godages, Colombo, 2006

Microenterprise Development: A Strategy for Poverty Alleviation & Employment Creation in the Third World: The Way Out of the World Bank and the IMF Stranglehold, Sarasavi, Colombo, 1977

Non-Formal Education Theory and Practice at Comilla, The Bangladesh Academy for Rural Development, Comilla, Bangladesh, 1984

Administering Rural Development in the Third World, The University Press, Dhaka, Bangladesh, 1983

The Vidane's Daughter (A Novel), Sarasavi, Colombo, 1998

TABLE OF CONTENTS

PREFACE

I T IS SAD TO realize that Democracy is actually on hold in the USA, the mother of democracies in the modern world. In the 2000 US Presidential Election it was clear to everyone that the election was stolen. In fact it can even be said that the words used by Vice President Al Gore when he conceded defeat implies that the election was actually stolen:

> Let there be no doubt, while I strongly disagree with the Court's decision I accept it ... for the sake of our unity as a people and the strength of our democracy I offer my concession.

Al Gore did not to want to cause harassment and embarrassment to a new Administration. He gracefully bowed out. In the case of the 2004 Presidential Election the election was closed by the machine count. To all it appeared a clear victory. But as the media one by one came up with gory details of the malfunctioning of electronic voting machines it became clear to the discerning that the victory had no basis or foundation whatsoever. The people of America were robbed without their being aware that they were robbed. It was done in a very subtle manner, so subtle that many would not even imagine that anything wrong had really happened.

The 2000 US Presidential Election was stolen by the Judiciary and this has left an indelible blot on the history of democracy in the United States of America, the longest standing democracy in the world. Many are the books that deal with this, including mine: *The Administrative Bungling that Hijacked the 2000 US Presidential Election*, The University Press of America, 2004.

To add insult to injury the 2004 US Presidential Election has been also stolen, not by the non-counting of votes but by the electronic voting machines that did the tabulation and totaling of ballots. This book is on how the electronic voting machines accomplished that task. It has to be conceded that the machines did that task efficiently and swiftly. It documents how the electronic voting machines that were used have been consistently malfunctioning in the tabulation of votes, adding or subtracting not tens or hundreds of votes, but millions at times and this has been happening well before 2000. The incidence of malfunctioning has been very high and even this study has had to deal with select instances.

In the 2000 US Presidential Election there was the possibility of manually counting all the votes in the deadlock that ensued. In fact Al Gore should have asked for a full recount of all ballots in all Florida 39 counties but he instead asked for a recount of select counties, where he felt that he had more votes and which he thought would have enabled him to clinch a victory. But the correct method would have been to have counted all the votes by hand to find the real victor. Harking back to my days of election work in Sri Lanka, I can state without any doubt and hesitation that if I or any of my colleagues as Assistant Returning Officers in charge of a count, faced a similar situation, we would have invariably ordered a full recount of all votes given the stark discrepancies that were brought to light. There would be no question whatsoever about such a decision. It would have been administratively as well as morally wrong to have decided otherwise. If we did not decide it that way, we would have been grilled by the Judiciary on an election petition and would have had to face censure and demotion.

My experience of counting ballots tells me that such a State wide recount could have been done. It is easier to attend to such a task in the USA. I have been a doctoral student in Michigan for two years and in recent years have spent months in travel, living off and on for long periods as my sons are domiciled in USA and am familiar with the working of public institutions in the USA. In my estimate there are hordes of officers in public institutions that can with a modicum of training easily handle a count of ballots and do it efficiently and accurate in a very responsible manner.

In the case of the 2004 US Presidential Election, things were otherwise. Even if one wanted to count the votes manually, there was no paper trail in many areas and therefore a manual recount was not possible. The totals coughed up by the electronic voting machines were final. It was actually an irrevocable decision. Democracy was automatically subverted and overridden.

Democracy can only be ensured by the proper administration of the election process. The Constitution lays down the rules and regulations, the Congress and when called upon, the Judiciary has a role to play when there are problems with interpretation and subversion. Within all this the administration is the live part of politics, where the rules embedded in the sacred Constitution gets ordained into an action process. If this administration process becomes tainted with partisanship, if the cogs in that process do not mesh properly and systematically due to inefficiency or manipulation-happening incidentally or premeditatedly then it happens to be Democracy that is sacrificed.

As an elections administrator I am sad to say that in both the Presidential Elections of 2000 and 2004 it was the fault of the administration of the election process that enabled the miscreants to steal the elections. This book on the 2004 Presidential Election as well as my earlier book: *The Administrative Bungling that Hijacked the 2000 US Presidential Election* truly highlight how the administrative process was hijacked. It is hoped that the administration of the election process is given due regard and recognition in both academic political studies in Universities as well as in actual practice at County and State levels. It is only such sustained action that can ensure that no future election is ever stolen.

My teacher, Professor George H. Axinn of the Michigan State University in his Foreword to my book on the 2000 US Presidential Election wrote:

Like many other US citizens who voted in the Presidential Election of the year 2000 I was surprised, disappointed, shocked, amazed and finally bewildered by the unanticipated series of events.

The outcome of the 2004 US Presidential Election too surprised, shocked and bewildered many, when the malfunctioning of the voting machinery was revealed. In Congress Ms. Norton said that the USA should have a failsafe method. The only failsafe method is a paper trail which can be manually counted in case there is a discrepancy. It is sad that the political authorities fail to realize this fact. The Republicans are not interested because it is a well known fact that the electronic voting machine manufacturers are supporters of the Republicans. But it is perhaps a travesty of fate that the Democrats too do not in one voice call for a paper trail before the 2008 US Presidential Election is held. If that is not done one cannot be certain whether a third election will also be stolen again. This time it may not be by the non-counting of votes, not be through the electronic voting machines; no one can fathom how the election will be stolen again. Democracy has to win and for that there has to be eternal vigilance.

Garvin Karunaratne
Center for Global Poverty Alleviation
2 Broadlands Avenue, London SW16 1NA, THE UK

ACKNOWLEDGEMENTS

AM DEEPLY INDEBTED TO the Journalists, Professors and other Experts who have voiced themselves exposing the irregularities found in the 2004 US Presidential Election. Their efforts would not have seen the light of day if not for the active and enthusiastic cooperation shown by the Editors of the various Newspapers and Journals who have deemed it their duty to publish the writings and to the publishers of the various books. It is hoped that their effort to save the democratic process is not in vain. However I take responsibility for the ideas I have opined.

I am also in deep debt to members of my family, my wife Bimba for appreciating and supporting my working on the cause for World Development and Democracy, my sons Arjuna and Kanchana for technical support and advice on computers and my son Hiran for his very searching comments.

I am also indebted to the staff of BookSurge-Amazon.com, especially to Nina Brauer, Angela Johnson and Aaron Volker for the enthusiasm shown in publishing this book.

Garvin Karunaratne
gamkga@aol.com
November 1, 2007

1

INTRODUCTION

Americans aren't really voting. Machines are.
Call it faking Democracy. (Lynn Landes)

THE 2004 U.S. PRESIDENTIAL ELECTION was hijacked not by administrative bungling but by inadequacies in the system of electronic voting that was characteristic of the total election. This does not mean that there was no administrative bungling, but the evidence indicates beyond all doubt that it was the manipulation done by the electronic voting machines in the counting of the ballots that tipped the scales for the victory of George W. Bush.

The more one reads the opinions of computer experts and professionals on computer malfunctioning, the more one goes through the numerous instances where computers malfunctioned in recording, tabulating and totaling the results of voting in many elections where electronic voting machines have been used, the more one comes to know the dubious connections that exist between the manufacturers of voting machines and the officers handling elections, the more one gropes and finds out the undue partisanship shown by officials at the helm of handling elections, the more one finds that optical scanning machines recorded undue gains in actual votes for Bush far above what was expected by the Exit Polls, as compared to the complete correlation of Exit Poll results and actual results in precincts where the electronic voting machines were not of the optical scanning type, the more one finds instances where the electronic voting machines coughed up more ballots cast than the actual number that did vote in the election, the more is one convinced that something radically wrong occurred in the recording, tabulation and totaling of votes in the 2004 Presidential Election. This is not a surmise or a guess. It is something that did really take place. This is contrary to the views of the National Research Committee on Elections and Voting, a panel comprising top election scholars who after due deliberation found no conclusive proof of a stolen election in the 2004 Presidential vote. However this Committee itself concluded that there was *pervasive breakdowns in elections administration and oversight, making it impossible to definitely put theories and accusations of fraud to rest. (Hill: 2006:4)* In my opinion this comment is a frank admission that something was definitely wrong, a wrong that they could not directly pinpoint as otherwise such a negative comment was not warranted. What I find is that in the case of the 2004 US Presidential Election it was very clearly the manner in which the electronic voting machines recorded, tabulated and totaled the ballots that decided the election in favor of George W. Bush. It has to be pointed out that the findings and recommendations of the National Research Committee on Elections and Voting is not based on true facts, as would be revealed in this Study.

In *The Analysis of An Electronic Voting System*, Tadayoshi Kohno of the Department of Computer Science at the University of California at San Diego, Adam Stubblefield and, Aviel D. Rubin of The Information Security Institute at John Hopkins University and Dan S. Wallach of the Department of Computer Science of Rice University, all academic specialists in computer science, working in institutes of great repute and standing, state that:

> Our analysis shows that this voting system is far below even the most minimal security standards applicable in other contexts. We identify several problems including unauthorized privilege escalation, incorrect use of cryptography, vulnerability to network threats and poor software development processes... We conclude that this voting system is unsuitable for use in a general election. (Kohno et al.: 2004:3)

The above Analysis is dated February 27, 2004. With what happened at the 2004 US Presidential Election, nine months later in November 2004, it is sad that this authoritative opinion was not heeded. The evidence at hand points out to that fact that what took place was a deliberate and well thought out plan of stealing an election, which took the non-Republican parties completely off guard.

Having personally handled the administration of elections as the chief officer in charge of the total election process in a major District in Sri Lanka and having worked as an officer in charge of voting precincts many times and later working in charge of holding elections as an Assistant Returning Officer at District level in two parliamentary elections, I am more than inquisitive as to how the democratic election process has become a prey firstly to riches and the availability of money if not for which luxury one cannot run for public office as a Mayor of a City, a Governor of a State, as a Senator or a Congressman or a President. In fact it is well known that the lack of sufficient funds precluded the former Democratic Vice President Albert Gore from coming forward as the Democratic candidate for the Presidency in the 2004 election. Secondly, the Democratic process has fallen prey to manipulation by partisan elected bureaucrats.

The 2000 US Presidential Election was hijacked by the Judiciary by stopping the counting of ballots. It is to the credit of the democratic forces in the USA that despite skirmishes and irregularities the ballots were available for counting. The election took on the form of a duel between the Judiciary at the State level which was decreeing that the ballots should be counted while the Federal Judiciary decided to stop the counts, thus eating into the democratic rights of the citizens which the Judiciary was actually avowed to safeguard and uphold.

It is observed that there have been serious irregularities in both the 2000 as well as the 2004 Presidential Elections. In both elections the irregularities reported are legion in number. Former US President Jimmy Carter had to say of the 2000 Presidential Election that *some basic international requirements for a fair election are missing.* (Carter, 2004) He was particularly referring to partisanship shown by election officials and to irregularities in voting and counting procedures. The elections administration in the US is riddled with corrupt practices due to the fact that the elections are handled by partisan officials at the State level as well as at the County level. True democracy has deemed that the chief officers who play a major part in the elections — the County Supervisors of Elections, the Secretary of State etc., are all elected and in these elections political party affiliations play a major role. However it has so happened that these officials who have been appointed to such important positions have not been educated enough to understand the importance of functioning in a non-partisan manner in their official duties. It is sad that they have failed to understand the fundamental fact that once elevated to high office they serve the entire nation and not a segment of it.

In the case of the 2000 US Presidential Election, it was the vacillation and the inefficiency of the election officials that led to the mishandling of the Florida Election, moving the election into the realm of the Judiciary. As pointed out by me, *Any General Election once the election process commences is on a time scale that cannot be varied and it is up to the administrator in charge to make the decisions carefully and within the specified time. Of course every decision that the administrator makes can subsequently be questioned in the judicial courts but it is the duty of the administrator to make the decisions in a carefully reasoned manner to ensure that the election is brought to its proper end and not delay till the aggrieved concerned parties decide to call for redress from the Judiciary. (Karunaratne: 2004:1)*

The Election need not have strayed into the judiciary if the officials in charge of the election had handled their duties reasonably and efficiently. This is a statement that I can make without any reservation because having handled the count of two parliamentary elections in Sri Lanka, in 1960 and 1965, I was aware that if I, had arrived at any unreasonable decisions or had handled the administration of the counting of ballots in any manner where I could be considered partisan or if I had vacillated or unduly delayed arriving at decisions in due time, if while being in charge of a count I had vanished and if none knew where I had gone or for what purpose, the process of decision making would be open to suspicion and would have invariably ended in the Supreme Court of Sri Lanka.

In the case of the 2004 Presidential Election there was administrative bungling but facts point out to how the laxity in certification of the electronic voting machinery and the very high incidence of malfunctioning of the electronic voting machinery led to a situation where unfortunately, *it is not who votes that counts but who counts that votes.* This amounts to a negation of democracy and it is worse when one realizes that the USA is the most long standing democracy—truly, the mother of democracies.

The experience in the use of voting machines had clearly shown many irregularities in results, pointing to the one fact that that electronic voting systems could be manipulated. It is not an experience privy to the 2000 year. Instances are replete where the voting machinery had malfunctioned as far back as two decades or more. This malfunction happens in the software because it is the software that codes how the markings on the ballot paper are recorded on the hard drive of the computer and the memory cards that store the ballot voting and tabulate it to a total. In short a programmer can code in such a manner that a vote cast for X can be counted as two or more votes in tabulating and similarly the codes can be programmed for any vote cast for Y to be reduced by 100% when tabulating or totaling. Even to a novice in computers it is clear that a change can be done when the software is written, but sad to say none of the Democrats who had a say in the US Presidential Election 2004 had the foresight to see through this pervasive loophole through which the election was to be hijacked a second time.

It was expected by the Democrats that in any corner of the country irregularities or complications in the counting of ballots could arise like in the pattern of what happened in the 2000 US Presidential Election and plane loads of lawyers, professionals and activists were available at hand to move at a moment's notice. *Kerry's lawyers were only trained to look for voter intimidation and similar incidents, not possible computerized fraud.* (The Command Post: *Nov 18, 2004:3*). The local Democrats could then be supported to face the situation. It is sad that the possibility of software manipulations was ignored. This did not even occur to them because otherwise the Democrats could have insisted on a paper trail in every State and also the proper certification of electronic voting machines.

In any electronic voting system it is absolutely necessary that there should be a fool-proof method that would enable a review of the counting, to verify whether the totals coughed up by the voting machines are correct. Sad to say there were no paper trails in the voting machines that were used in the States of Ohio and Florida. These were key States with massive populations and any erroneous counting in these States could easily tip the scales in the election.

A final word is due of the method of study that has been used by me in writing this book. Instances of serious administrative bungling and the malfunctioning of the electronic voting machines that have been used in the USA, both before the 2004 US Presidential Election as well as in the 2004 Presidential Election are really legion. The instances are so many that if I attempt to compile every instance and draw conclusions the book will easily run into thousands of pages and will never be a feasible publication. Instead it will be an academic compilation that will hog the shelves of a few research universities never completely read by anyone. It is therefore decided to draw specific glaring instances of irregularities in a representative manner and draw conclusions.

Further, the concentration in this Book is on the 2004 US Presidential Election. The 2000 US Presidential Election is comprehensively covered in my book: *The Administrative Bungling that Hijacked the 2000 U.S. Presidential Election* (The University Press of America) and it is not hoped to repeat its contents unless such references are essential for the presentation of a complete picture in this study.

2

ADMINISTRATIVE BUNGLING CONTINUES
UNABATED

*Effective administration is necessary in order
for democracy to actually work.
(Professor George H. Axinn)*

T IS OBSERVED THAT administrative bungling that was so
characteristic of the 2000 U.S. Presidential Election, highlighted
in my book, *The Administrative Bungling that Hijacked the 2000
U.S. Presidential Election* continued in the 2004 U.S. Presidential
Election too. In fact even the very insignificant newspaper, the
Michigan State University's *State News* states:

*The 2000 Presidential Election left an indelible stain on our
Electoral College ... Four years later, we've all learned that
not much has changed. A recent study by the Associated
Press found that most Americans were pleased with the
decisive nature of last week's election. We respectfully
disagree. Last week's election didn't have a decisive plotline.
Ohio was perilously close to becoming the new Florida. The*

only decisiveness about the Presidential Election was Kerry's decision to accept defeat rather than drag his country and the Electoral College through the same process it went through four years ago. (11-9-04)

Actually, it is important to note that even if Kerry had contested the declared result, there was absolutely no method of establishing who really won because there was no system of independent recording on paper that could have corrected the outcomes coughed up by the electronic voting machines. The verdict given by the machines had to be invariably accepted for the purposes of declaring the winner.

Though the administrative lapses that were noticed in the 2000 US Presidential Election were laid bare in detail by me in my earlier quoted book, and also repeatedly, in no mean manner by enlightened academia, political activists, writers and journalists, no definite remedial action had been taken, except in very rare instances. As stated in *Business Week* (June 14, 2004):

The underlying problems that led to the Presidential election crisis still exist and could stretch on for years.

Errors do happen due to the negligence of officialdom. It can happen incidentally due to inefficiency. It can also happen premeditatedly. Instructions and procedures are laid down in order to ensure that any election is conducted in a systematic, reasonable and fair manner and it is up to the elections administration to adhere. Even where it is clear that certain officials had been negligent no punitive action has been taken. In my own experience as an elections administrator in Sri Lanka, if the negligence and inefficiency that was so characteristic of the 2000 US Presidential Election both at the State level and in certain definite instances at the County level had happened in the case of the elections where I was in charge of the administration, heads would have rolled and definite action would have been invariably taken to clean the Augean stables. I am certain that the officials would not only have faced dismissal from their substantive posts in the government service but they would also have faced incarceration. Well that was in Sri Lanka, a poor insignificant country, a country that could be dubbed by some to be a banana republic. What should we expect of the mother of democracies-the enlightened USA, of all countries, is the crucial question at issue?

INEFFICIENCY AT THE HELM OF
ADMINISTRATION

THE ELECTIONS ADMINISTRATION IN any State comes directly under the Secretary of State. In the 2000 US Presidential Election it was found that the Secretary of State at Florida, Katherine Harris had played a nefarious and dubious role. I would repeat what I wrote in my book on the 2000 US Presidential Election:

> To my thinking it was the duty of the Florida Secretary of State to have decided on the standard for the manual count and also to have ensured that such standard was followed on a uniform basis by all county canvassing boards. This is purely an administrative decision that should have actually been decided by the Florida Secretary of State in consultation with the county canvassing boards, well before the aggrieved parties went to the Judiciary. It was precisely the failure of the Florida Secretary of State to make a decision and implement it on a uniform basis across all Florida counties that directly led to the crisis. (Karunaratne: 2004:35)

Even after the inefficiency and the vacillation in arriving at decisions and the partisanship of such officials had been amply proved, it is found that they have got off scot free without any censure or punishment. The Florida Secretary of State has belatedly stated that she was only functioning at the State level and that the problems were caused by the failure of the County Election Supervisors. The very fact that the Secretary of State admits that there was a failure is very important. However as far as responsibility is concerned The Secretary of State cannot absolve herself of responsibility as the State Department of Elections comes directly under the Office of the Secretary of State.

On the contrary it is sad to note that officials at the helm-at the level of Secretary of State who have tried to clean the Augean stables have had to face censure. Kevin Shelley, the Secretary of State at California, battled to implement a statewide system to prevent voter fraud. He was forced out of his job. However, though he had to bow out, the effort was not in vain. In California a State Law in 2004 required all electronic voting machines to produce paper records of every vote cast at any election. In view of the fact that the electronic voting machinery had not been altered as at May 2006, in the June 6, 2006 Primary Election in Alameda County, California, it was decided that paper ballots replace the touch screen voter systems that had been used for the past five years. It has been openly admitted that the flaws in the voting machines had not yet been corrected. *The reason is that our current touch screens do not have the printer attachment that allows you to verify your vote as required by a new State Law. (Alameda County Registrar of Voters: Sample Ballot & Voter Information Pamphlet: 2006)*

In the Santa Clara County in California, a new paper record system called the VeriVote was introduced in 2006. This was done by an attachment to the Sequoia touch screen voting machines. This enables a voter to print and review a paper record of the ballot. *The record on a roll like a cash register receipt scrolls through a window on the VeriVote box. The voter then selects either 'make changes' or 'cast ballot'. Once the ballot is cast the screen will show 'vote recorded' and the paper record will scroll out of view... The County will store the paper records for 22 months. (San Jose MercuryNews: May12, 2006)*

Instances of negligence and inefficiency that can be proved to have been done with a malicious intent to be partisan in the election have been many. In connection with the elections in Colorado's Seventh Congressional District, when a recount had to be done, *complicating the count were eleven directives and instructions from the Colorado Secretary of State which local elections officials complained were confusing and contradictory. (Goldstein: 2003:96)* It was no surprise that it took as long as five weeks after the election to declare the winner. It is also important to note that the instructions that caused confusion emanated from no less a person than the highest official in the State-the Secretary of State.

When the officials at the helm have been acting in a very nefarious partisan manner, when they should actually have conducted themselves in a non-partisan exemplary manner, it is important to note that they cannot expect the officers at the County level to be non-partisan, perfect and efficient in their work.

THE ADMINISTRATION OF VOTING PRECINCTS

A S STATED IN MY book on the 2000 US Presidential Election, there should be standing orders of instructions written according to the Laws-Federal and State which would clearly lay down the procedures that have to be invariably followed at all voting precincts. There should be a chain of command where it is mandatory for particular statutory officers like the Secretary of State, the Commissioner of Elections and the County Supervisors of Elections to take action. Further there should be procedures that have to be mandatorily followed when an irregularity is noticed. Irregularities that can occur have to be identified and classified; the action that has to taken should be laid down also specifying the name and designation of the officer who has to take action.

It is necessary that each Voting Precinct should be manned by very experienced officers and the presiding officer or the officer in charge should be an officer who can arrive at decisions and have effective supervision over the conduct of the election at the precinct center. This was often not the case which warranted Representative, Mrs. Jones to state in the Congress:

We should require that those working at the polling booth to be fairly compensated, adequately educated and sufficiently supported such that the job importance will be elevated (Congressional Record-House: January 6, 2005; H121)

Conducting elections should be accorded serious concern and the staff should be responsible and experienced. As commented by me, *the lack of a specialist staff in elections has impeded the elections administration both at the County as well as at the State level in the 2000 US Presidential Election. (Karunaratne: 2004:96)* It is important to note that this serious shortcoming has not yet been corrected.

VOTER INTIMIDATION & HARASSMENT

NTIMIDATION AND HARASSMENT OF voters has been a common feature in the 2000 US Presidential Election as well as earlier. Even the Police, the supreme law enforcement agency had played a very dubious and partisan role. (Karunaratne: 2004: 90–93)

In the 2004 Presidential Election too, many are the instances where people or political parties had to evoke the Judicial Courts for justice due to voter suppression. This need not have happened if the administration of the Voting Precincts had been done well. Many are the instances where the Republicans had alleged that the Democrats were intimidating voters and vice versa.

The Orlando Sentinel reports that *A Seminole County Circuit Court Judge issues an injunction to block Democrats from intimidating Republican poll watchers who seek to challenge voters. (Orlando Sentinel: Wisconsin www.demos-usa)*

Fond du Lac Reporter from Wisconsin states that *Long lines in Wisconsin deterred voters. One of the causes of the delay was harassment from challengers. (Fond du Lac: www.demos-usa)*

It is reported that in Seminole County, Florida, there were complaints of voter intimidation.

(WKMG6FL; www.demos-usa)

The Wichta Eagle reports that there was voter intimidation at a Benedict College Precinct in South Carolina (*The Wichita Eagle: www.demos-usa.*)

In Michigan, *Mercury News* reported that the GOP sued the City for allegedly throwing out challengers from precincts. (*Mercury News: www.demos-usa*)

Harassment by intent is evident when one sees the very large amounts of voters that had been challenged or *announced that it plans to challenge 37,000 Milwaukee voters* (GM *Today: www. demos-usa*)

It is reported that about 6,000 challenges were made in Florida and in Ohio at least 125,000 challenges were made *(New York Times: Nov 7, 2004)*.

In Pennsylvania it was reported by *The Philadelphia Inquirer that the Republicans are poised to contest as many as 10,000 registered voters. (The Philadelphia Inquirer: www.demos-usa.)* Here too, the motive appears to make the polling unworkable. The Republicans were perhaps trying to avoid the democrats from voting.

In these instances the motives of challengers appear to have been to create confusion, chaos and an unworkable situation in the voting precinct because every challenge needs to be looked into by the polling staff, documented, reasoned decisions arrived at, appropriate action taken and documented. It is absurd for any one to target to have a particular number of challenges, as a challenge has to be done only when an impersonation is suspected or it is found that the person had voted earlier and no one can be certain as to how many such instances will actually occur within a particular precinct.

The task of a challenger is essentially to help the polling staff. The challengers are generally local people who know local residents and have the ability to point out that the prospective voter is either impersonating or has voted earlier. The polling staff is in most cases comprised of outsiders who do not know the local people. Thus the role of a challenger is complementary to the polling staff and can be said to be functioning in an essential process in the holding of a free and fair election.

It is reported that challengers were harassed:

> At the University of Wisconsin Madison Campus ... the Republican's lone monitor found himself outnumbered by democratic counterparts who shouted him down each time he questioned a voter. (Wall Street Journal, Nov 3, 2004)

If a challenger was shouted at, it reflects the basic fact that the administration of the voting precinct was weak and inefficient. Shouting of any sort, intimidation of any sort should not have been tolerated. In such an instance it is the duty of the officer in charge of the voting precinct to bring the person who shouts to order and order him out if he/she persists. The police should be asked to move out such persons. In fact the Police have the ability and it is their duty to arrest such persons for disorderly conduct and even charge them for obstructing the polling. The decision to pursue legal action against any challenger for disrupting the election by his disorderly conduct or impersonation has to be decided by the officer in charge of the voting precinct.

One can hark back to the 2000 US Presidential Election when the Miami Dade Canvassing Board was reviewing 10,750 uncounted ballots a riot was organized by House Republican Whip Tom Delay ... causing a riot that frightened the Canvassing Board. (Karunaratne: 2004:92) This effectively stopped the recount. It is important to note that the persons who staged that riot held very responsible office. Actually the persons responsible for creating the riot should have been prosecuted, which was not done. It is only firm action taken in such instances that can ensure the proper conduct of an election.

As pointed out earlier, the right to challenge a voter is an essential part of the democratic right of citizens and the administration at the voting precinct should have seen to it that a challenger can exercise his right. A challenger who feels that the voter is an impersonator should have the right to challenge but it would be up to the officer in charge of the voting precinct to look into the bona fides of the voter, decide whether the voter is an imposter and either hand over a ballot in case the voter proves identity or to hand over him/her to the legal officers for prosecution in case it is proved that the person was attempting to impersonate. It is my opinion that if firm action had been taken in the case of the first few challengers such action would have served as a deterrent for both impersonators to come forward as well as for challengers to challenge. The ruling of a Judge to ban challengers could be construed to deny the democratic rights of the people. It may have been that the judge was forced to give the ruling because the officers at the voting precinct were allowing the challengers to run riot. It is up to the official in charge of the voting precinct to ensure that the challengers are allowed to challenge in a reasonable manner.

In my experience as a Presiding Officer at parliamentary elections in Sri Lanka I had to deal with challengers personally. Challenging can become a problem only if there is inefficiency on the part of the elections staff, especially the officer in charge of the polling precinct.

2.4

LONG LINES OF VOTERS AT POLLING PRECINCTS

ONG LINES AT POLLING Stations was widely reported in the 2000 U.S. Presidential Election. Long Lines can be caused by inadequate staff and/or inadequate voting machines. These are factors that could have been easily rectified through administrative decisions. It is simply posting adequate staff and ensuring that adequate voting machinery is available. The number of voters is definitely known and the administrators in charge at the County level should have posted adequate staff to attend to the voting. The number of voting machines should also have been adequate with a few standby machines to be used if there is a breakdown. In each County there should also have been a pool of extra machines at a central place that could be transported to any voting precinct at short notice. These are extremely simple administrative matters and having served as an administrator for over four decades it really beats me as to why such a lapse could have happened.

Many are the instances of delays and long lines of voters at the voting precincts in the 2004 US Presidential Election.

The Charlotte Observer stated that in some precincts in the State of Carolina, the scene *was like Soviet bread stores, with lines spilling onto sidewalks and wrapping around buildings* (Gumbel: 2005:290)

Long Lines were reported in Chicago. (*Chicago Sun Times*: *www.demos-usa.*)

Long Lines were reported from Florida. (Drinkard: *USAToday*: *Nov 18, 2004*)

Long lines were reported in Franklin and Knox Counties in Ohio. The Lines were so long and this amounted to be a formidable problem which even made a Federal Judge to order to provide paper ballots to all voters still in line in Franklin or Knox Counties (*Columbus Despatch*: *www.demos-usa*)

In Ohio, generally long lines resulting in delays up to 2 to 3 hours was very common while there were isolated instances where the delay was 7 to 9 hours. (*NewYorkTimes*: Nov 07, 2004)

Long Lines were reported in New York City. (*NewYorkSun*: *www.demos-usa.Org/page196.cfm*: AA)

At Gambio, Ohio, long lines were reported. This was due to the fact that there were only two voting machines to serve 1,170 voters. The voting had to continue till 4 A.M. A voter who went to vote at 1.30 P.M. voted only at 11 P.M. (*Drinkard: USA Today*: *Nov 18, 2004*)

In Kenyon College precinct in Gambio, Ohio, there were long lines and voters took 6 to 11 hours to vote. (*Wikipedia: 2004 US.... Controversies: Ohio*)

Long Lines were reported at poll sites in Pennsylvania. (*The Patriot News(PA):www.demos-usa.org /page196.cfm*)

In Oberlin, students that came to vote had to wait hours, varying from five to ten hours to vote. (*Solon: Nov 7, 2004*)

The House Judiciary Committee in a fifteen page letter asked the Ohio Secretary of State Kenneth Blackwell for explanations re widespread irregularities re the elections in Ohio. This included a call for explanation for the fact that in the Franklin County long lines were found in predominantly Democratic precincts. (*House Judiciary Committee Dec 4, 04*) (1)

It is alleged that planning to create long lines is a strategy by which voting was selectively suppressed. The simple method used in this case is to administer the voting process in a manner that would deter voters. This has generally happened in areas where the population was well known to be principally Democratic. Long lines meant that people took hours-at times six or more which will dissuade people to vote. Many who cannot wait that long would go away without voting. This will amount to a negation of a person's right to vote.

It has to be pointed out that the prevalence of long lines at voting precincts is a serious situation and it appears that in many such instances, the officials at the helm have premeditatedly subverted the Democratic process by the simple method of not providing sufficient voting machines or by inadequate staffing.

OPENING AND CLOSING OF VOTING PRECINCTS

N CONNECTION WITH THE 2000 US Presidential Election, I wrote *There should actually be the same number of voting hours throughout the country with definitely fixed hours for voting. These should not be varied at the whims and fancies of the State Elections Division or the County Canvassing Boards (Karunaratne: 2004:70)*

It has been found that the opening and closing of voting precincts has not been uniformly done in the 2004 US Presidential Election. It should never happen that a voting precinct is closed during voting time. This practice has caused confusion among voters.

In Florida, it is reported that *hundreds left a Hialeah polling place this morning without voting after election workers shut down the polls for 45 minutes shortly after opening time. (Miami Herald: www.demos-usa)*

In Michigan it is reported that NAACP filed a lawsuit to extend the hours of polling until 8 p.m. *(Mercury News: www.demos-usa)*

The Pittsburgh Channel reported that: *a Pennsylvania Judge has extended the voting deadline for provisional ballots until 9.30 p.m. in Allegheny County.(The Pittsburgh Channel. www.demos-usa)*

Ensuring that the opening and closing times of voting precincts is done properly in a uniform manner is a simple and straightforward administrative task, which has been neglected. This reflects very badly on the elections administration.

INADEQUATE POLL WORKERS

S HORTAGES IN POLL WORKERS reflect the inefficiency of the administration. This is also an extremely simple administrative task.

The *Newsnet 5.com* highlights that in Ohio, the shortage of poll workers have caused a near standstill in a Cleveland Polling place. Some voters have left without voting, while others have waited 5 hours to vote. (*Newsnet5.com:www.demos-usa*)

The problem lies in the fact that in the USA the elections departments depend mainly on volunteer staff to man the elections. Training in elections law and procedures is something that cannot be done in a few hours to volunteers that can be casually found. The irresponsibility of volunteers becomes repeatedly evident. In a particular instance, poll workers who were posted at some garages were feeling cold and had used unused ballots to make fires to keep warm. (*Wall Street Journal*: Nov 2, 2004) Each ballot paper is an important document and has to be securely kept. This instance shows the irresponsibility of the elections staff and entrusting an election to persons of this caliber can only end in sheer disaster.

Having had personal face to face dealings with many officers in American Universities, in County and State administrative organizations, during my doctoral studies in Michigan and in my stay-for years with my sons domiciled in California, travelling all over in almost every State in the US, I have been thoroughly impressed with the ability and efficiency displayed by officers and I can vouch for their responsibility and fairness. It is sad that this officialdom was not used to run the elections. In Sri Lanka we take the cream of officialdom from Government and Local Government institutions-all handpicked, to ensure that the elections are efficiently conducted. In Third World countries like Sri Lanka and India poll workers are selected from among the government and local government permanent staff that has a good record of work. They are thoroughly trained. If any poll worker were to act either in a partisan manner or colludes with any irregularity they are very likely to pay for it by being punished which could leave a permanent black mark on their employment record. Entrusting volunteers and third grade workers which poll duties is not a method by which a fair and clean election can ever be held.

TRAINING OF POLL WORKERS

RAINING HAS TO BE invariably provided to poll workers. They have to be aware of their duties and be also conversant with the law. While at least 2 million poll workers were required the Election Assistance Commission has said that only 1.4 million poll workers had been trained. A further problem was that *for every three poll workers trained only two show up on Election Day. (Assoc Press via Yahoo News: www.demos-usa)* It follows that over half of the poll workers had not been trained.

WNBC reports of poll workers in New York City. polling precincts in that they were poorly trained. *(WNBC: www. demos-usa)*

In the Colorado's Seventh Congressional District when a mandatory recount had to be done 535 uncounted votes came to light due to the fact that the poll workers did not know how to handle the new voting machines. (Goldstein: 2003:96)

In precincts in Pennsylvania and Indian lands in New Mexico, *election watchers reported widespread confusion among precinct workers over how the (provisional)ballots are supposed to be used. (Culmers et.al: Wall Street Journal: Nov 3, 2004)* This had to happen because even before the 2000 election the poor quality of elections staff had been amply commented upon but no remedial action taken. The Wall Street Journal, stated in 2000: that local counties, *hire poorly trained staff for the massive job of recounting and counting a total of more than 100 million votes for president as well millions more for lower offices. (Dec15, 2000)* Actually this makes a mockery of the elections administration in a developed country like the US. *(Karunaratne : 2004:71)*

William Welsh states of the inexperienced poll workers: *In some places poll workers were unaware they needed to boot up the e voting machines before voting began. In other places election officials forgot to plug in the machines, which caused them to shut down when their batteries ran out. In a few cases poll workers inadvertently activated the procedures for closing voting rather than opening it on the machines. Because the carefully monitored machines cannot be restarted once the closing process starts new machines had to be sent to the field. (Welsh: 2004)*

Proper training in election work is essential to ensure that a fair and clean election is held. This is an area that has been totally neglected in the USA. This is due to the practice of enlisting volunteers for election duty. Instead, the cadres for election duty should be found from among veteran officers working in local government and other public institutions. This will enable their training to be undertaken on a definite basis. The use of such trained officers who hold definite posts in the public sector will ensure that the elections are conducted in a systematic and non-partisan manner.

MAINTENANCE OF KEY ELECTION RECORDS

K EY ELECTION RECORDS LIKE the used ballot papers, records-logs maintained by the officers in charge of the polling precincts have to be securely kept. This is an aspect that has been totally neglected in the USA. Instances where key records had not been securely handled are frequent.

In the 2000 US Presidential Election there was a glaring instance in the Seminole and Martin Counties of Florida. TheElection Supervisors-Sandra Gourd of Seminole County and Peggy Robbins of Martin County had allowed Republicans to add missing voter identification numbers to absentee ballots after they had already been filled. As stated by me,

> It is actually not known as to what endorsements or entries were made by the Republican officials or even whether the ballots were substituted. Once the absentee ballot is received by the County Canvassing Board it is the guarded property of the canvassing board and under no circumstances should they be released to anyone unless on a Court Order. (Karunaratne: 2004:62)

The serious nature of the lapse in not ensuring the security of the ballot papers and instead, releasing them to Republican officers is that the entries made by the Republican officers enabled the validation of these ballots which can be said to have tipped the Florida election in favor of Bush. Normally when duly filled absentee ballot papers are received action is taken at the Registrar's office to check the signatures. This indicates the fact that the authenticity has to be cleared and under no circumstances can such ballots be released to anyone. The fact remains that no punitive action was meted out to the officers responsible for this lapse. This reflects the laxity on the part of the elections administration.

In the 2004 US Presidential Election, in view of the problems that were clearly evident in the State of Ohio, Votewatch, a San Francisco based activist group had requested copies of key election records from the 88 Counties in Ohio. All records were not available in a single county. In the case of ten Counties less than half the critically important items such as signature rosters, ballot accounting reports, accumulated total reports, were missing. (Gumbel: 2005:292)

This reflects a serious situation. It need not be over emphasized that the non availability of such critically important documents indicates the lackadaisical and casual manner in which the election had been conducted. In my experience as the Assistant Returning Officer in the case of the electorates where I was in charge in Sri Lanka general elections in 1960 and 1965, if the critical reports were missing I would have taken the precinct officer in charge and the record keepers of the elections department to task.

It has already been mentioned how poll workers had resorted to burning ballot papers to keep warm. This shows that ballot papers had not been handled with sufficient care. In this case the ballot papers were unused. However even unused ballot papers have to be handled with security and care.

VOTER'S REGISTERS

ONGRESS MANDATED THE CREATION of computerized Statewide registration. However it was found that *more than 80% States have failed to comply. (USA Today: Editorial: Nov 10, 2004)*

Registration problems have been common. This is due to the fact that the registration of voters was not closed on a particular date and voter registration was allowed at the voting precincts for the issue of provisional ballots. Despite the fact that this is a simple task that can easily be solved many registration problems were reported in the 2004 US Presidential Election.

Voter registration has to be done on a strict basis. It is the laxity in voter registration that leads to the necessity for issuing provisional ballots. In Ohio as much as 155,000 had to use provisional ballots because of *registration mix ups or other problems. (USA Today, Editorial: Nov 10, 2004)* This indicates the fact that the elections process in Ohio had serious failings.

At New Jersey, registration problems led to hearings at the Rutgers Constitutional Law Clinic., making Judges allowing over 200 would be voters to cast their provisional ballots. *(Miami Herald: www.demos-usa)* This is a task that could have been handled by an efficient administration at the voting precinct. This need not have gone to the Judiciary.

Capital News9, reported of problems in Albany, New York: hundreds are waiting in line at the Board of Elections in Albany to sort out errors that are preventing them from voting(*CapitalNews9: www.demos-usa*)

It has been an accepted practice for voters to be registered when they proceed to get driver's licenses. In North Texas, dozens of people who had registered to vote when filling for driver's licenses found that their names were not on the register, (*NBCDallasFortWorth:www.Demos-usa*) The people had to prove residency by presenting a driver's license, utility bill or other form of documentation. Any voter who does not provide proof of residency will be asked to take an oath before being allowed to vote. The Registers were not authentic and names had been added on without proper verification of residence which led to challenges. This explains why there was so much of confusion at the 2004 Presidential election.

The House Judiciary Committee in a fifteen page letter asked the Ohio Secretary of State Kenneth Blackwell for explanations re widespread irregularities the elections in Ohio and this included the fact that in voter registration in Perry County, many of the voter's files did not have the signatures of the voters. (*House Judiciary Committee: Dec4, 2004*) The signatures should have been there and in its absence should have been rejected. This reflects badly on the elections administration at the county level, where the voter registration is done and the register is maintained.

There is also the practice of registering voters by their political party that can lead to complications. In the 2004 Election Lawyers for the Republicans accused Democrats of registering fictitious characters as voters (*Becker & Finkell: October 31, 2004*) Republicans registering voters have been accused of registering only Republicans. This situation of allowing political party activists to attend to voter registration is inimical for the conduct of a fair election.

Further, the practice of paying workers per registration is also likely to involve fictitious names with the idea of claiming money. It is reported that in Colorado there were 719 fraudulent forms submitted for registration. (*Deborah Sherman: 2004-10-11*)

Voter registration is an important task that cannot be left in the hands of political activists whose aim is to ensure that their party wins at the elections. It is a task that has to be attended to by the permanent elections staff in an essentially non-partisan manner.

In the USA voter registration processes are also handled by private companies. Entrusting voter registration to private companies should not be done because the voter's register is an important document and the validity of the entire election rests on the Register's authenticity. There are reports of voter registration companies whose *bosses trashed registration forms filled out by Democratic voters because they only wanted to sign up Republican voters.* Russel, who worked for Voters Outreach of America states that he witnessed his bosses ripping up registration forms that had been filled by Democrats. They were thrown in the trash. Russel when questioned stated that he does not know the number of registration forms so thrown out but adds that the number is high because this company worked in Las Vegas for over two months. (*Wikipedia: 2004. Controversy: 39, 40*)

The voter registration system in the USA has come in for criticism by the International Monitors. In their words: *The United States is also nearly unique in lacking a unified vote registration system or national identity card.* (*IntHeraldTribune: 2004/11/03*)

Voter registration is simple once the procedures have been put into place. It would be ideal to have a single countrywide norm and effectively administer the registration of votes., (Karunaratne: 2004:52) Inadequacies in voter registration is a major failure that should not have happened.

2.10

FELONS

FELONS ARE NOT ENTITLED to vote. In most States after serving the sentences felons are immediately restored onto the voters register. However, Florida is among half a dozen States that do not automatically restore civic rights to felons who have completed their punitive sentences. In fact it is reported that in Florida no proper administrative action had been taken to restore the voting rights of felons that had served their sentences. The details can be said to amount to intentional action by the Governor. *The Miami Herald* reports that the problem of felons being purged from the Voters Registers came up *despite Governor Jeb Bush's pledge in 2001 to improve Florida's clemency system; ex felons continue to have the restoration of their voting rights blocked. Even though the number of cases has grown from about 6,000 to 65,000 during Bush's time in office the Governor has refused to hire more staff members to handle the backlog and has denied rights to more than 85% of applicants.* (The Miami Herald: www.demos-usa)

It is reported that *the State of Florida purged over 90,000 people from their list of eligible voters under the guise that they were felons. In fact almost none of the disenfranchised were felons ... but almost all were blacks and democrats.* (Wikipedia: 2004. Controversy: 23)

The *Des Moines Register* reports that in Iowa, *Many Iowans mistakenly purged from the Voter Rolls after been incorrectly identified as felons. (Des Moines Register: www.demos-usa.)*

In Colorado, Democrats had complained that an attempt was made to remove 6,000 felons on the instructions of Donetta Davidson, the Secretary of State, *despite a US Federal Law that prohibits eliminating a voter's rights within 90 days of an election to give time for the voter to protest. (Wikipedia: 2004 . Controversy: 52)*

This administrative malaise reflects very badly on the Department of Elections. It is petty action that, when neglected, amount to a total negation of democracy in the long run.

BALLOT PAPERS

THE PROPER HANDLING OF an election hinges greatly on ballot papers and therefore the design, printing and handling of ballot papers has to be attended to with great care, importance and security. Currently there are ballot papers belonging to the political parties. These are partisan ballot papers, i.e. ballot papers belonging to the main political parties and non-partisan ballot papers for the others. The voter's political party is encoded on his/her voter card. Different ballot papers for the same election can be confusing and complications have only to be expected.

It is found that the design of ballot papers is not properly defined-*different voters can be presented with different ballots depending on the party affiliations ... if an attacker(hacker) changes the party affiliations of the candidates then he may succeed in forcing the voters to view and vote on erroneous ballots. (Wikipedia: 2004. Controversy: 13)*

It is absurd to allow different ballot papers for different political parties. All ballot papers for any particular election shown to the voter on the electronic voting machines and on the paper trail should all conform to a particular design and pattern that has been approved by the Elections Department. There should be no variation whatsoever.

2.12

MISPRINTED & CONFUSING BALLOT PAPERS

B ALLOT PAPERS ARE PRINTED at printing establishments and also turned out by the electronic voting machines. Ballot papers should be printed under high security to ensure that there are no counterfeits. In view of the fact that ballot papers can also be printed/produced by electronic voting machines that have been proved to have malfunctioned, it is imperative that action has to be taken to ensure that the software used in such machines are also not tampered with.

Confusing Ballot Papers are due to faulty designs, which can be interpreted differently by the voters, either due to misprints or due to faulty alignment. The effect that the Butterfly ballot paper had in Palm Beach County, Florida in the 2000 US Presidential Election is well known. While the seriousness of what happened in the 2000 Presidential Election by the use of the butterfly ballot should have acted as an eye opener to election officials country wide, it is sad to realize that no lesson has been learned.

VerifiedVoting.org reports that in Chicago and Cook County *a cluttered facing-page layout that confronted voters deciding whether to retain sitting Cook County Circuit Court Judges. It was so perplexing that scores of people complained that they voted for Judge A when they meant to vote against Judge B. (VerifiedVoting. org: March 16, 2004)*

The *New York Times* states: *Ohio continues to use confusing butterfly ballots in parts of the State. (New YorkTimes:www. demos-usa.)*

In the 2004 Presidential Election, too misprinted ballot papers were reported in California *(LA Times:www.demos-usa)* It is reported that the confusing nature of this ballot paper necessitated two ballots being given to voters. This led to a problem in counting because it was possible that *the duplicate ballots may not have been removed (wftv.com:www.demos-usa)* This could have been easily avoided if the second ballot paper was given only when the earlier ballot paper had been surrendered. This should have been covered under standing orders.

Enforcing standards in printing ballot papers is a simple task that has so far been ignored. *The onus of responsibility for the confusion caused by the faulty design of ballots falls squarely on the elections administration both at the county and at the state level. (Karunaratne: 2004:61)*

Professor Rebecca Mercuri said in 2001 that *There are no required standards for voting displays, so computer ballots can be constructed to be as confusing (or more) than the butterfly used in Florida giving advantage to some candidates over others. (Mercuri: 2001:1)*

It is not a difficult task to avoid confusing and misprinted ballot papers. The Secretary of State and his Elections Department should advise the County Supervisors on this and ensure that all ballot papers to be used at any election conform to certain definite specifications. It is really a sad reflection on the Elections Departments at the State level to have not yet ensured that ballot papers are not confusing, cannot be misinterpreted and that no proper care was taken. This inefficiency persists to this day.

PROPER CARE OF BALLOT PAPERS & VOTING MACHINERY

THE NECESSITY FOR THE proper security and care of ballot papers and voting machinery and related equipment need not be over emphasized.

The fact that the administration was weak in the 2004 US Presidential Election is reflected in incidents where no proper care has been taken.

In Alameda County in California, a News Reporter Kim Zetter who reported for poll worker training found that *her fellow trainees were entrusted with keys and combination numbers enabling them to get access to the country's Diebold machines ... There would have been plenty of time for any mischief since the DRE terminals and the memory cards used to operate them were left sitting in polling stations days before the election.* (Gumbel: 2005: 265) This indicates that no proper care had been taken in handling the electronic voting machines that print the ballot papers. The care of ballot papers is a simple and uncomplicated task. Keys, combinations and memory cards should be held in the custody of the officer in charge of the voting precinct.

In Miami County, Ohio, it is reported that *the audit log information for the November 2, 2004 election was completely missing.* (Peckarsky: Nov 6, 2006)

The Stealing of six Diebold tabulation machines and a touch screen voting terminal on June 10, 2002 from a Ramada Inn in Macon, where they had been used for training is important in that these were never traced. (*Gumbel: 2005:236*) A Voting machine being totally lost is serious in that its technology can get into the hands of a hacker and be misused. This could have enabled the hacker to rig all similar machines even on a nationwide basis.

Several hundred wet ballots were found in Seminole County, Florida and these were rejected by the voting machines. (WKMF6FL: www.demos-usa)

The allegations made by Black Box Voting Inc. (BBV) against the Supervisor of Elections, Volusia County included *that portions or all of the voting machine tapes for 59 precincts out of a total of 159 precincts are missing*

BBV reported finding the following when they attended Volusia County:

A trash bag found in the warehouse ... found to contain official voting records,

Records supplied by County Officials printed the day before without signatures ... were checked against the originals and differed from the original voting records,

Official voting tapes were found designated for shredding,

Tapes in the official garbage, designated for shredding. contained at least one of the missing tapes

(Volusia County Fraud Investigation, Nov 16, 2004 in Wikipedia : 2004. US Voting Machines)

Voting machines have to be stored securely at all times, and there has to be very strict standards of security adhered to on a mandatory basis. This would apply to all accessories of the voting machines too. The Administration of Elections has been extremely weak in taking care of voting machines. There have been many instances where the voting machines were not properly stored and/or adequate measures not taken to ensure their total security.

2.14

ABSENTEE BALLOTS

THERE WERE TWO MAJOR instances of serious irregularities in the 2000 US Presidential Election where Republicans were allowed to take charge of thousands of absentee ballots on which Republicans had cast their vote, make alterations or additions which would have under normal election law tantamount to a cancellation of the election as well as incarceration of the officers who were responsible for the proper care of the ballots. (*Karunaratne: 2004:61–63*) Despite this gross mistake, which should have taught a lesson, in the case of the 2004 US Presidential Election too, the handling of absentee ballots had posed a major problem.

It is necessary that absentee ballot papers have to be posted in time. There have been numerous instances of delays and of mishandling the mail.

In the 2004 US Presidential Election too, there have been delays in sending out Absentee Ballots. Sending out absentee ballots is an administrative task which is simple and this reflects inefficiency that has to be deplored.

In Ohio, a federal judge had to be invoked to intervene and order that voters who had requested absentee ballots but did not receive them had to be given provisional ballots. (*ToledoBlade: www.demos-usa*)

The *Philadelphia Inquirer* reported that in Pennsylvania, a federal judge had ruled that *election officials must wait to count more than 12,000 absentee ballots until after a hearing tomorrow at 9.30 p.m. (PhiladelphiaInquirer: www.demos-usa)*

It is a ridiculous situation for a judge to make such an order when counting ballots is something that has to be done invariably. This shows the inadequacy of the administrative instructions re counting of votes.

In Broward County, Florida, *The ACLU filed a lawsuit Tuesday asking Broward County election officials to extend the absentee ballot deadline until November 12 since thousands of voters did not receive their absentee ballots in time. (Brandenton Herald: www. demos-usa)*

The Broward County population is heavily democratic and this raises the issue whether this was purposely done.

The very vast numbers of absentee ballots misplaced or lost, undelivered can easily upset an election result. In Florida, in the Broward County 58,000 absentee ballots posted were never received by the voters. *(Houston Chronicle: www.demos-usa.).* While the US Postal Service had followed strict procedures and supervision over employees it is possible that there may have been a few lapses. *An e-mail from a U.S. Postal Service District Manager indicates that Postal Service employees may be mishandling ballots. (SouthFloridaSunSentinel: www.demos-usa)*

The Broward County Board of Elections was expected to send up to 14,000 duplicate absentee ballots within a day after it was discovered earlier that tens of thousands of Floridians never received a ballot.

In Ohio there were 62,513 absentee ballots. Dr Werner Lange points out that when the precinct poll books were checked for absentee voters and the number of actual absentee voters was compared to the certified number of absentee votes, *there was an inflated difference in nearly every precinct of the five communities examined ... the 106 precincts of these five Ohio communities, about 39% of all precincts in Trumbull County netted a total of 580 absentee votes for which there were no absentee voters identified in the poll books. (Lange: Dec 12, 2004)* These discrepancies indicate the careless manner in which absentee ballots had been handled.

It is reported that New Mexico's absentee ballots were posted nine days later than scheduled because many Counties had to reprint them with Nader's name. The Supreme Court had decided that Nader's name should be included only on 17th September. *(Cummings: Nov 3, 2004)*

These glaring instances indicate that the elections administration was inefficient and lax in attending to absentee ballots. Lessons had not been learned even after the catastrophic misadventures so characteristic of the 2000 US Presidential Election.

It is important to note that the handling of absentee ballots has been outsourced-given to outsiders while in actuality it should be done by responsible officers under the close scrutiny and supervision of the County Election Supervisor. The absentee ballots have to be sorted out when received and this is a task that cannot be subsequently checked. (Black BoxVoting: 194) It is reported that *The Prison release documents of Jeff Dean (Diebold) state that he was employed by Postal Services Inc (PSI) the company which counted these votes. The job was later sub contracted to Diebold's mail division, which he ran before passing it on to John Elder, another felon he met in prison. (Wikipedia: 2004 ... Vote Suppression)*

It is ridiculous to outsource the handling of ballots to outside companies. This reflects how the election process is being administered without any concern for security.

It was necessary to provide the main political parties advance copies of approved absentee ballots. (Philadelphia Inquirer: www.demos-usa) To my mind, issuing ballot papers whether it be absentee ballots or provisional ballots is left to the Elections Administration and it is not necessary to provide advanced copies of absentee ballots to each political party.

PROVISIONAL BALLOTS

THE PROBLEM WITH PROVISIONAL ballots is that the elections officials have to ensure that the voters who cast them are registered voters. This is time consuming and in the event of large numbers, becomes an enormous task.

In the 2004 US Presidential Election, many are the instances where precincts ran short of provisional ballots. This is a sad reflection on the elections administration.

WTAVTv reports that in Pennsylvania, there was a shortage of provisional ballots in 22 precincts and that it was necessary for the voters to use the voting machines instead. This even led to an application to Courts to extend the voting time. (*WTAETV: www. demos-usa*)This is due to the fact that the number of provisional ballots given to each precinct had been insufficient.

Randall Tousaint, a volunteer who attended to voter registration and monitor polling at an eastern Georgian County states that in a poll precinct at Savannah State University 25 provisional ballots that were available had been used by 11 a.m. and thereafter no provisional ballots could be issued. This happened to be an area where the main population was black. (*Gross: Plain Dealer: Nov 06, 2004*) What is clear is that the number of provisional ballot papers that had to be given to each voting precinct had not been carefully decided.

Precincts in Pennsylvania and in Indian Lands in New Mexico ran out of provisional ballots. (Culmers et.al. *Wall Street Journal:* Nov 3, 2004)

KDKA2Pittsburgh reported that 100 local precincts in Western Pennsylvania ran out of provisional ballots by 3 p.m. (KDKA2Pittsburgh: www.demos-usa)

In the 2004 US Presidential Election, the Ohio Secretary of State Kenneth Blackwell initially insisted that provisional voting can be done only at the proper polling station. However this was contravened by a US District Court Judge ruling that Blackwell's ruling *would violate Federal Law if it prevents voters from casting provisional ballots if they are in the right county.* Instead of complying with this judicial ruling, Blackwell *issued new guidelines that Democratic lawyers say failed to select the court ruling. Poll workers are now supposed to attempt to turn away provisional voters if they are at the wrong precinct and give them a telephone number to locate the correct one. If the voters insist, the workers are to accept the ballot while warning that it will not be counted.* (ShawnMcCarthy: 2004)

It is a sad reflection on the elections administration that people of the stature of Secretaries of State have resorted to make such questionable rulings.

In Ohio a total of 155,337 provisional votes were received. Many irregularities were reported in their counting. For instance, the Secretary of State Blackwell had ruled that provisional ballots should not be counted for 11 days, which did cause some ballots to remain uncounted. Many of the provisional ballots had been from areas where Kerry supporters were heavily represented. When Kerry conceded 155,428 provisional ballots, 92,672 "spoilt" ballots, additional overseas ballots and some remaining absentee ballots were uncounted. (Fitrakis: Nov 07, 2004) Kerry lost Ohio by only 136,483 votes. For a victory to be correctly decided these ballots should have been counted.

A Provisional ballot was normally accepted if the voter's name and address were correct and if the voter's signature matched the signature in the County's data base. However in Cuyahoga County in Ohio a new ruling had been implemented which required the date of birth to be written on the packet. (*Lonewacko Blog: Nov 11, 2004*). A similar order not to accept the provisional ballots was made by Kenneth Blackwell the Secretary of State at Ohio, but this ruling was later cancelled. (*All Experts: 2004 ...*) The insistence of such new provisions makes a mockery of counting. The specifications should be the same for the entire USA and should not be varied.

If the Voters Registers had been compiled properly, there would be no need for any provisional ballots. The problems involved with the issue of provisional ballots can easily be obviated by concentrating on the voter's registers.

It is important to note that two elections officials in Cuyahoga County, Elections Coordinator Jacqueline Maiden and Ballot Manager Kathleen Dreamer were found guilty of negligent misconduct in the 2004 election. They were accused of *secretly reviewing pre selected ballots before a public recount on Dec 16, 2004. They worked behind closed doors for three days to pick ballots they knew would not cause discrepancies when checked by hand, prosecutors said (Int.Herald Tribune: AP: Jan 24, 2007)*

SPREAD OF MISINFORMATION : SABOTAGE

THE ADMINISTRATION HAS TO take responsibility for the proper holding of the election. This includes total surveillance over the political process, beginning at the Primary Stage through to the holding of the election and the counting process. Throughout this period, action should have been taken to stop any miscreants from creating confusion. It would be irregular and also illegal for confusion to be wantonly created. Such wanton measures were commonly reported in the period of the 2004 Presidential Election.

The Washington Post tells of what happened throughout the country:

> Dirty tricks targeting voters continue across the country. Students in Florida are unknowingly having their party registrations switched. Fliers distributed in Pennsylvania are telling people that the Republicans should vote on November 2 and Democrats on November 3 due to the expectation of a higher voter turnout. Blacks in Milwaukee are being told that they will be sent to prison if they vote in the Presidential election after having already voted in another election this year. (TheWashingtonPost: www.demos-usa)

This is confirmed by Dennis Kucinich, Democratic Representative in Ohio, who states of misinformation that happened in his State:

> *Dirty tricks occurred across the State, including phony letters from Boards of Elections telling people that their registration through some democratic activist groups was invalid and that Kerry voters were to report on Wednesday because of massive voter turnout. Phone calls to voters giving them erroneous polling information were also common. (Wikipedia: 2004 US Presidential.)*

The *Washington Post* reports that in Cuyahoga County, unknown people visited homes offering to collect and deliver absentee ballots. In the words of Jane Platten, of the County Board of Elections *we've never seen anything like this before ... there seems to be a concerted effort to give voters misinformation.* (Becker & Finkell: Washington Post: Oct 31, 2004)

ABC12 reported that in Michigan *voters receive blatantly false electioneering phone calls.* (ABC12: www.demos-usa)

The *Capital Times* tells of a GOP flier distributed in Wisconsin: *a Flier distributed on Friday to six University of Wisconsin residence halls urges students to be at the polling place of their choice. In Wisconsin, voters must cast their ballots at the polling place (where they are on the Voter's Register) in order to be counted.* (The CapitalTimes: www.demos-usa)

In Alleghaney County in Pittsburg, a bogus but official in appearance flier was found stating that due to immense voter turn out it had been decided that Republicans should vote on Tuesday and that Democrats should vote on Thursday. (*Washington Post:* Oct 31, 2004)

Voters had been given the incorrect instructions re their voting precincts. Michell Hargett said that *she found a flyer on her door sending her to the wrong precinct, miles away from the precinct where she is registered. (AFP Features: Nov 02, 2004)*

Another method of sabotage is to have either inadequate staff or inadequate voting machines which can cause long lines of voters, make voters who do not have the time to tarry hours in a queue go away without voting. While this can be due to inefficiency it can also be done premeditatedly to avoid people voting.

It was found that in the Cuyahoga County, Ohio, electronic voting machines were malfunctioning more in black areas. In detail, *of the 82 precincts for which voters reported that one or more voting machines were not working, the vast majority was in neighbourhoods where over 75% of the population were blacks. In one precinct, 7 out of 17 voting machines were not working. In another 3 of 9 voting machines were not working. In yet another 2 of 3 voting machines were not working. In two precincts all the voting machines were not working for a significant period during the day. Multiple reports of voting machines which highlighted a vote for Bush when Kerry's button was pressed. (Wikipedia: 2004 .... Vote Suppression)*

Spreading misinformation with the idea of misleading voters is tantamount to the subversion of democracy. It appears that the spread of misinformation was not incidental. Instead it appears to have been premeditated and done intentionally to create a problem or to provide an advantage to a particular candidate. It reflects on the election administration for such misinformation to have been allowed to continue and for allowing such machine shortages in particular areas. Further in each instance the elections administration should have reported the matter to the Police and it was the duty of the Police to investigate and arrest any person who had aided, abetted or accomplished such illegal tasks.

On the whole it is clear that like in the case of the 2000 US Presidential Election in the 2004 US Presidential Election too the administration has failed to ensure that a fair election was held. This is a sad situation to fail in implementing proper procedures which could have ensured the prevalence of democracy.

3

THE VOTING MACHINES FIASCO

*The essence of democracy is the confidence of the electorate
in the accuracy of the voting methods and the
fairness of voting procedures
(John Conyers, Republican Congressman from Michigan)*

*If you want to win an election, Just Control The Voting
Machines (Thom Hartman)*

A MAJOR FACTOR IN THE debate between the use of electronic voting machines and the use of paper ballots lies in that while in the use of paper ballots the count will be decided by the actual counting of ballots, in the case of electronic voting the technicians and engineers who program software, construct and maintain the machines play a key role in bringing about the outcome of the election. The fact remains that the programmer can program for the votes cast to be tabulated and totaled in any manner he/she desires. This can be verified and checked only by a technically qualified computer software expert. Bruce Schneier , one of the reputed US cryptographers is of the opinion that *no computer voting system be adopted unless it also provides a physical paper ballot perused by the voter and used for recount and verification. (Quoted by Mercuri: 2001:1)*

It is important that the physical paper verified by the voter is preserved for any audit recount and not stored in the voting machine or on a store card-CD or disk, because such storage systems can be electronically programmed to submit different results. Thus it is imperative that there should be a paper trail of the voting which is kept in safe custody separate from the machines. This Paper Trail, in other words-a paper documentation of the voting is an essential factor that would alone enable the verification of the results of a vote. In the words of Rebecca Mercuri, *electronic balloting systems without individual printouts for examination by the voters do not provide an independent audit trail.* (Mercuri: 2001:1) In the USA as a whole, California set the standard in connection with voting machines having a paper trail. Kevin Shelley, the California Secretary of State in November 2003 by law mandated that all DRE systems should have *a voter verified paper audit trail in time for the 2006 mid term election.* (Gumbel: 2005:265)

Despite the fact that the necessity for an independent paper trail is absolutely necessary to ensure that democracy prevails-in that the candidate who gets most votes is elected, it is found that there has been foot-dragging and delays in implementing such a decision. The San Jose Mercury News tells us of the blatant fact that the majority of ballots cast at the 2004 Presidential Election cannot be checked even by the voter. It reads:

> A printer-one that lets voters verify on paper the electronic votes they cast on a touch screen voting machine-will remedy much of the controversy surrounding electronic voting. Unfortunately such a device won't be at the polls in Santa Clara County or most of America in this presidential election.

This comes of great importance because in writing the software for an electronic voting machine one can code it in such a manner that a voter can see that he has voted for a particular person on the screen to enable the voter to be satisfied, but in totaling a different code can be used to distort the voting. In short the second coding to be used in totaling could order the computer to add every second vote for Mr. X to the total of Mr. Y. or any other combination. Professor Rebecca Mercury says:

Any programmer can write code that displays one thing on a screen, records something else, and prints yet another result. There is no known way to ensure that this is not happening inside of a voting system. (Wikipedia: 2004. Controversy: 125)

Anyhow a foolproof system is very necessary as has been pointed out by the *San Jose Mercury News*:

We've said for twenty months now that many of the suspicions surrounding the touch screens would evaporate if the voting companies produced and Congress required a voter verified paper trail for electronic voting. They haven't, despite revelations of security gaps and coding weaknesses of touch screen systems.

It gave a warning for the forthcoming 2004 presidential election:

Without a paper copy that voters have seen and confirmed there will be no way to fully resolve charges of manipulation, fraud and error with touch screen machines. (SanJose MercuryNews: 15-08-2004)

It is even reported that *the Secretary of State, Blackwell is seeking to make any vote not cast by an electronic machine illegal. (Columbus Despatch: www.demos-pusa)* It is a moot question as to whether a Secretary of State has the power to make such a decision. Paper ballots are to be used only when there was a problem with the voting machinery and such a decision can be said to be totally undemocratic leading eventually to the disenfranchisement of voters.

COUNTING OF BALLOTS BY ELECTRONIC VOTING MACHINES

I T HAS BEEN ESTIMATED that *with the highly insecure and error prone touch screen voting machines (which will process 28.9% of all votes this year) a huge threat still remains-computerized ballot scanners. They will count 57.6% of all votes cast including absentee ballots. (Landes: Two Voting Companies: 4/27/2004)*

Actually over 80% of all ballots voted are tabulated and totaled by electronic voting machines.

In the words of Professor Rebecca Mercuri,

Fully electronic systems do not provide anyway that the voter can truly verify that the ballot cast corresponds to that being recorded, transmitted or tabulated. (Mercuri: 2001:1), "Cindy Cohn," the Legal Director at "Electronic Frontier Foundation" *opines that the average vending machine is more secure than the Diebold Code. (Zetter: Aug 12, 2003)*

While in the counting by hand errors can occur, such errors can be minimized and even totally obviated by careful counting. Even in technologically advanced countries like Germany ballots are counted by hand. It takes a long time but accuracy can be assured. In my experience as the officer in charge of counting ballots in two electorates in Sri Lanka, I can vouch for the fact that it was possible to avoid all errors and declare a winner. In the counts done under my supervision I can vouch for the fact that it was 100% fool proof. The method was firstly to separate the ballots by the candidate for whom the votes were cast and then making them into bundles of one hundred, recounting the bundles again and again, In one electorate where I presided over the counting (Walapone Electorate in Nuwara Eliya District) the winning margin was low and I, with a handpicked staff counted every ballot thrice in the presence of the two candidates and a few of their representatives. The total ballots counted in each were in the region of forty to fifty thousand and the recounts took around six extra hours. What is required in the case of the USA is to divide a County to sections-by precincts and have a hand picked responsible staff-not volunteers, to count the ballots manually in the event of a recount. As an experienced administrator, with experience of working in five countries including the UK I can vouch for the fact that this is very feasible and the one and only method of ensuring that democracy prevails. This is also very necessary considering the experience of using electronic voting machines in the USA, where despite it being the most technologically developed country there appears no end in sight to the irregularities that have come to light in counting ballots. The only conclusion that can be arrived at is to the effect that the voting machines can malfunction and can never be relied upon.

Many-legion are the instances where electronic voting machines have malfunctioned. It is impossible to make a note of every instance where the voting machines have malfunctioned. However the glaring instances are given below to enable conclusions to be drawn.

EXPERIENCE OF MALFUNCTIONING VOTING MACHINERY IN THE PRE 2000 PERIOD

A S FAR BACK AS 1970, a report by the accountancy group Price Waterhouse stated:

> It is possible to have instructions in computer memory to call in special procedures from core, tape or disc files to create results, other than those anticipated. (Gumbel: 2005:190)

Such warnings had been ignored by the authorities.

In 1971, in the City Assembly Election in Las Vegas, Nevada, Democrat Arthur Espinoza was declared the winner, but after unrecorded votes were counted, his opponent Republican Hal Smith was elected. (The Las Vegas Review Journal: Nov 30, 1994)

In 1980, Michael Shamos, the Pennsylvania voting equipment examiner stated of the Votomatic voting machines that they could easily be tampered with

> Shamos showed how arbitrary numbers could be entered into the machine's counters and how with just a little practice and no tools of any kind, in less than a minute a voter could disassemble the ballot page mechanism and rearrange. (Gumbel: 2005:190)

In 1985 in a Mayoral Election in Dallas, Mayor Starke Taylor was trailing behind Max Goldblatt when there was a power failure. When power was restored a few minutes later Taylor was in the lead. Further odd things were reported-*in one District 425 Goldblatt votes recorded at 10.55 P.M. on election night simply vanished. In another area known to favor Mayor Taylor, the number of ballots reported jumped from 295 to 547. in the words of the Attorney General's Office: The electronic voting system in use lacks adequate security features to provide any assurances of the absence of fraud.* The Assistant Attorney General Robert L. Lemens wrote: *this office has found that it will be difficult to demonstrate to the complainants that Texas elections are free from fraud and thereby free local election officers from suspicion.* (Quoted by Gumbel: 2005:194)

In 1986, in Georgia, in the State Senator election in District 48, the voting machines declared that the Democrat Don Peevey had lost. However, it was found that a computerized voting program had miscounted and when corrected, Peevey was declared the winner. It is important to note the peculiar procedure adopted for the correction: *When the count finished around 1 A.M. they (the elections board) walked into a room and shut the door ... When they came out they said, "Mr Peevey, you won". That was it. They never explained. (The Atlanta Journal Sept 3, 1998)* The decision had been arrived at without any observation by the candidates. In actuality, the candidates should have been present at the count. The Elections Board should have taken in the candidates in or if they were not present, their representatives, to see how the count was conducted. In Sri Lankan election counting, the observations of the candidates are noted in writing, authenticated with their signatures. It is my opinion that in any recount there should be transparency for the decision to be accepted by the parties concerned.

In November 1993, in Kane County Illinois, the results for a dozen election results were found incomplete and in one case a local referendum proposal was announced lost and later corrected. This is said to have happened due to the voting machines omitting the results of eight precincts. (*Chicago Tribune: April 04, 2003*) Omitting the votes of even a single precinct is something that reflects inefficiency. Checks should be carried out repeatedly to ensure that the results of all precincts are included. Omitting the results of eight precincts reflects gross inefficiency.

In November 1996, in the McLennon County, Texas in the Republican Primary Runoff in one precinct the tally was over 8000 votes when only 500 votes were cast.. It's a mystery, as said by Linda Lewis, the Elections Administrator. (*Dallas Morning News, April 13, 1996*)

In a Seminole Nation Election in August 1997, in Oklahoma the voting machine count declared the wrong winners. This happened because the computer had counted the absentee ballots twice. (*Newsbytes News Network: August 05, 1997*) This also shows that coding can cause mistakes in tabulation.

In the Salt Lake City Election in 1998, 1413 votes failed to show up in the count. When these votes were found the decision of the election was reversed. (*The Salt Lake Tribune: June 25, 1998*)

In the April 1998, School Bond Referendum in Orange County California, at first the Registrar announced that the bond issue was lost and later corrected it and reported that the bond had won. This was attributed to a programmer reversing the yes and no answers in the software used in counting the votes. (*Harris, 2003:12*) & (*Newsbytes: 04-24-1998*) This clearly indicates that as stated earlier, the coding can change the results declared by voting machines.

In Prima County, Arizona, in the 1998 general election, the computers recorded no votes for 24 precincts, but voter rolls showed thousands had voted at those polling stations (*Arizona Daily Star: 11 Nov 1998*)

In a November 1999 in a Council seat election in Onondaga County, New York, a software programming error had given too many absentee votes to Republican Faulkner pointing out a victory for him, but when the absentee ballots were properly re-checked, Democrat Lytel was declared elected. (*Newsbytes News Network: April 22, 1999*)

In November 2000, in counting the ballots for Precinct 20, at Glenwood Springs, Colorado, it was found that the ES&S software did not function and a corrected chip had to be obtained to tabulate the results. (*City of Glenwood: 2000*)

MALFUNCTIONING VOTING MACHINES IN THE
2000 U.S. PRESIDENTIAL ELECTION

I N THE 2000 US Presidential Election it was found that *the voting machines in themselves have been eating into democracy, some having a failure rate of as much as 4% while other machines had a failure rate of 1%*. (*Karunaratne: 2004:114*)

In the 2000 US Presidential Election in Riverside, in the words of Andrew Gumbel,

> *A couple of hours after the polls closed the tabulation software overloaded and started deleting votes from the tallying system instead of adding them. Sequoia had to send in an emergency resuscitation team creating a delay of several hours.* (*Gumbel: 2005:226*)

A glaring instance of a malfunctioning voting machine in the 2000 Presidential Election was reported in the Wall Street Journal. This happened in an optical scan machine at Allamakee County, Iowa. When 300 ballots were fed, the machine reported 4 million votes. The County Auditor tried the machine again and it repeated 4 million. The fault was so vast that it could not be ignored. As Bill Roe, the County Auditor said, *We don't have 4 million voters in the State of Iowa*. The remedy that the manufacturer,

E.S. & S had was to get replacement equipment. (*WallStreetJournal, Nov 17, 2000*) The two facts that come to light are the magnitude of the malfunction which attracted national attention in a manner that could not be ignored and the inadequacy of the certification process. How was replacement equipment obtained? This new equipment could not have been certified. If it had been certified how was it in the possession of the manufacturer.

A Report from the Caltech-MIT Voting Technology Project estimated that *1.5 million presidential votes were not recorded in 2000 because of difficulties using voting equipment and that electronic machines have the second highest rate of unmarked uncounted and spoiled ballots in presidential, Senate and Governor elections over the last 12 years. (Infernal Press.Com: 11/29/2004)*

The evidence already quoted in this study suggests the high possibility of this statement being true.

MALFUNCTIONING VOTING MACHINES IN THE
2000–2004 PERIOD

IKE IN THE PERIOD before 2000, instances where electronic voting machines malfunctioned in the 2000 to 2004 period is legion.

In the 2002 General Election in Wayne County, North Carolina, the House District 11, the result was firstly miscounted. When the computer error was corrected 5,500 more votes showed up. (*The News & Observer: Nov 09, 2002*)

In the 2002, School Bond Issue in Gretna, Nebraska, the ES&S voting machines failed to tally the 'yes' votes and gave the impression that the vote was lost. But it was found when the votes were properly added that the vote was passed. (*Omaha World Herald: Nov 06, 2002*)

A worker in Diebold's Georgia Warehouse, states that *the company installed patches on its machine before the State's gubernatorial election that were never certified by independent testing authorities or cleared with Georgia elections officials.* (*Wikipedia: 2004. Controversy: 24*)

In November 2002, in Cherry Hill, New Jersey in the Mayor's election it was found that 46 machines malfunctioned and in addition 96% of the voting machines couldn't register votes for the mayor. It is important to note that all the machines concerned had been earlier tested and certified suitable. *(Newsday: "Voting Glitches" November 6, 2003)*

In the general election in November 2002, in Broward County *more that 100,000 votes mysteriously disappeared and were not recovered till the day after the election. The County's Deputy Election Supervisor blamed it on a minor software thing. (Gumbel: 2005:233)*

In the November 2002 Election in Fulton County of Georgia, *67 memory cards containing an unknown number of votes went missing, delaying certification of the results for ten days. In Dekalb County 10 memory cards went missing and were later recovered from terminals that had supposedly broken down and been taken out of service. (Gumbel: 2005:237)* A memory-card records a part or the entirety of the voting done in an electronic voting machine depending on how the voting is saved. It is a fundamental necessity to develop mandatory procedures regarding saving the votes cast on the memory-cards and the security that has to be ensured in the safe storage of the memory-cards.

In the November 2002 general election for Commissioner, in Scurry County, Texas a victory predicted by the machines for two Republican Commissioner candidates was overturned after a new computer chip was used to tally the votes. *(Houston Chronicle: Nov 08, 2002)*

In Bernalillio County, New Mexico, in the November 2002 Election 48,000 voters had voted on the un-auditable Sequoia Touch Screen computers, but of this number only 36,000 had been tallied. The President of Sequoia, Howard Cramer had apologized and mentioned of a similar irregularity in Clark County, Nevada. It was later that Sequoia technicians e mailed the correct results. *(Alberquerque Journal: Nov 19–2002)*

In the 2002, Clay County, Kansas, the Commissioner Primary election, the voting machines at first tallied that Jerry Mayo was losing, but a hand count proved that he won by 76% of the vote. *(Wichita Eagle: August 22, 2002)*

In the Florida State primary election of 2002, where Janet Reno Democrat was defeated by Jim McBride, Janet Reno noticed something suspicious in several precincts of Miami-Dade and Broward Counties and she called in Professor Rebecca Mercuri an expert in computers to investigate. In Mercury's words: *You would have places where there were over 1,300 (voters who had polled and there would be like one vote for governor,* which is highly ridiculous. E.S. & S. the manufacturer had sent a technician and Mercuri records what happened: *Basically E.S. & S. comes in and they've got some sort of tool they stick in some part of the machine and they pull some data out of it. How can you trust that?* *(Lefebvre: Dec 24, 2003)*

Though Janet Reno, the former US Attorney General strongly suspected that the eight thousand vote deficit separating her from Bill McBride was totally wrong there was no basis for her to make a complaint. (Gumbel: 2005:232)

In the Governor's Race in Alabama in 2002 the Democratic incumbent Siegelman appeared to have won by a narrow margin only to be undone by a sudden discovery of a computer glitch in rural Baldwin County. The County's probate judge in charge of elections had taken it upon himself in the dead of the night long after poll watchers and most of the staff had gone home.

Siegelman had accidentally been awarded seven thousand votes, too many-enough to tip the entire race to his Republican challenger ... County officials were distinctly vague about the cause of the error. (Gumbel: 2005:8)

Mark Kelley of ES&S when questioned is reported to have said *Something happened. I don't have enough intelligence to exactly identify what. (Mobile Register: Jan 28, 2003)* It is absurd for the tabulation and totaling of ballots to be done in secrecy. This lack of transparency should never have happened.

In the Boone County in Indiana, in the November 2003 election, the MicroVote Machines tallied a total of 144,000 votes when only 5352 votes were cast. *(Indianapolis Star: November9, 2003)* It is reported that *the Boone County Technology Director and a few Micro-Vote techs 'fixed the problem'.*

In Louisiana, in the elections for the Fifth Congressional District it was found *that the Republican candidate was ready to ask for a recount when it was discovered that he had lost by 974 votes, not by the previously reported 518 because an old Print-O-Matic machine had failed to print out its totals. (Goldstein: 2003:97)*

In August 2004, when the Sequoia machines were showed off its new paper trail system to lawmakers in California it was found that its Spanish language setting made arithmetical mistakes. (Gumbel: 2005:287)

Black Box Voting Org. stated that in King County Washington, they found serious irregularities regarding the conduct of the primary election six weeks ago. They *uncovered an internal audit log containing a three hour deletion on election night, 'trouble slips' revealing suspicious modem activity; and profound problems with security, including accidental disclosure of critically sensitive remote access information to poll workers, office personnel and even in a shocking blunder to Black Box Voting activists. (USA Today: E-Voting: Nov 03, 04)*

In the 2004 March Maryland Primary Election it was found that electronic voting machines in three counties did not record the voter's vote for Senator. Diebold officers are reported to have admitted that defective software had been used and also that the software had not been certified. (*Linda Schade: 2004*)

In the San Diego and Alameda Counties in California in the 2004 Primary it was found that *the devices used to generate individualized computer cards for each voter malfunctioned ... In San Diego County more than half the polling stations failed to open on time because the card device's batteries mysteriously drained. Finally the choice was between using equipment that had not been fully tested and approved or using no equipment at all. (Gumbel: 2005:267)*

The California State Voting Systems and Procedures Panel in April 2004 by an 8-0 vote *recommended that Shelley (the Secretary of State) cease the use of the machines saying that Texas based Diebold has performed poorly in California and its machines malfunctioned turning away many voters in San Diego County. (Jim Wasserman of Associated Press, Quoted by Landes: 4/27/2004).* This directly led to a decision to insist on a paper trail in California.

In the words of Marc Carrel, a Panel member, an Assistant Secretary of State: *The company has disenfranchised voters in California and undermined confidence in the new and developing technology of touch screen voting. (Wikipedia.org/wiki/2004:42)*

The Panel also documented that *573 of the 1038 polling places in San Diego County failed to open on time because Diebold voting machines malfunctioned. Voters were told to go elsewhere or come back. (Wikipedia.org/wiki/2004:42)*

A malfunction of the electronic voting machines in the Vote for Proposition 72 held in 2004 was recorded in San Diego County. The Secretary of State declared a victory for Proposition 72, which provided that large and medium sized employers should pay for at least 80% of their employee's health insurance. This was retracted by the Secretary of State and State officials blamed a faulty vote report from San Diego County. *"County officials said there was no mistake in their counting of votes for Proposition 72. Instead the glitch occurred because the State's and County's voting recording systems were out of sync ... we sent the right information but in the wrong order."* In detail, according to Sally McPherson, the County Registrar *the County's electronic recording system moved the results for Proposition 60A to the end of the reporting line-following Proposition 72 and submitted them to the State in that order ... The results were recorded out of sequence. (Conaughton: 2004)*

What is serious is that this type of sequencing mistake *was not unusual* and *happens all the time* in the words of County Registrar Sally McPherson. This irregularity is ascribed to the use of two different computer systems at the State Office and the County Office, resulting in interfacing problems. (Ibid)

The reporting by researchers at John Hopkins and Rice Universities highlighting *numerous programming faults and security vulnerability in the source code for Diebold's Accu Vote TS Voting Machines.* prompted Maryland Governor, Robert L. Erhlich Jr. to order a review. (Zetter: Aug 12, 2003)

Many are the glaring instances of malfunctioning voting machines that had been found before the 2004 Presidential Election, which should have indicated that some definite action should have been taken to ensure that such an irregularity would not occur in the 2004 US Presidential Election.

VOTING MACHINE MALFUNCTIONING IN THE
2004 US PRESIDENTIAL ELECTION

IGH INCIDENCE OF MALFUNCTIONING.
The instances of voting malpractices were in the tens
of thousands. In Florida alone there were over 4,369
incidents of voting problems reported to VoteProtect.Org. (*Bollyn:*
Dec 5, 2004)

The Associated Press states that *around 1,100 incidents of voting*
machine problems were reported yesterday. (Associated Press
YahooNews: www.demos-usa)

The *LATimes* reports of *malfunctioning DRE Machines.*
(*LATimes: www.demos-usa*)

The *Miami Herald* refers to the fact that there were
malfunctioning voting machines at precincts throughout the
State of Pennsylvania. (*MiamiHerald: www.demos-usa*)

The comprehensive investigative study undertaken by the
Democratic National Committee (DNC) found that *overall 28%*
of Ohio voters reported problems with their voting experience,
including ballot problems, locating their proper polling place and
or intimidation. (DNC: 10) This is a very high percentage.

In the 2004 US Presidential Election, the explanations asked
for by the House Judiciary Committee from the Secretary of

State Ohio, included an explanation as to why in Franklin County, *77 voting machines malfunctioned in the course of the day* (House Judiciary Committee: Dec 4, 2004)

The Miami Herald reports that there were *malfunctioning voting machines throughout the State of South Carolina. (MiamiHerald: www.demos-usa.) The Wichita Eagle* too confirms that electronic voting machines malfunctioned in the State and voters had to vote on paper ballots. *(The Wichita Eagle: www.demos-usa.)*

Verified Voting.org, a group founded by David Dill of Stanford University traced over 23,000 complaints on e voting on election day and 11,000 complaints belatedly; (*Gumbel: 2005:288*)

DIFFERENCE BETWEEN THE NUMBER OF VOTERS WHO SIGNED IN AND THE NUMBER OF BALLOTS COUNTED

In the 2004 US Presidential Election, a survey done by the *Miami Herald* found that in many precincts in Florida there was a difference between the number of voters who signed in and the number of ballots counted. When election officials were confronted the explanation forthcoming was that voters had failed to sign in. (*Chardy & Kidwell: Nov 27, 2004*)

In Perry County, the poll books examined after the election showed *more votes cast than actual voters voting.* There were instances where votes were counted twice by the machines which County officials blamed on computer errors.

(*"House Judiciary Committee Demands Explanation of Irregularities in Ohio": Black Box Voting: Ballot Tampering in the 21 st Century: Dec 4 04, (1)*

Of New Mexico, Michael Parenti points out that *in almost half of New Mexico's Counties more votes were reported than were recorded as been cast and tallies were consistently in Bush's favor.* (*Parenti: 2006*)

In Craven County, "*11,283 more votes for president than cast was reported.*" (*North Carolina's Ballot Blues: 12-04-2004*)

It is absurd to expect persons to be allowed to vote without their being signed in. Though this is reported as a minor discrepancy, in my experience as an election official who has officiated in charge of parliamentary elections in Sri Lanka this is a serious irregularity for which the elections officials in charge of the precinct should be held responsible. Working precincts with volunteers and such temporary and inexperienced persons as elections staff could be the reason for this type of irregularity. When working with inefficient, irresponsible, inexperienced and undependable staff, errors cannot be avoided. However this being due to a malfunction of the voting machines cannot be ruled out.

WRONG RESPONSE. WRONG CANDIDATES APPEAR ON TOUCH SCREENS.

USA Today (Nov 04, 04), reported that *several dozen voters in six States, particularly Florida, said the wrong candidates appeared on their touch screen machine's checkout screen ... In many cases voters said they intended to select John Kerry but when the computer asked them to verify the choice it showed them instead, opting for President Bush.* (USA Today: 11-04-04)

It was reported that in the Mahoning County, "*numerous voters reported problems with not being able to select Kerry on voting machines, which defaulted to Bush.*" (House Judiciary Committee: Dec 4, 04)

Fred Grimm states that a number of voters had told him that *when they voted on the ES&S system, they pushed Kerry and a vote for Bush popped up on their ballot summary.* (Grimm: Nov 09, 2004)

Verified Voting.org, found that of the over 23,000 complaints on e-voting on election day and 11,000 complaints belatedly, some were of the type that when they pressed the button for a particular candidate another's name appeared. No checking was possible. (Gumbel: 2005:288)

The Election Protection Coalition had received a total of 32 reports of touch screen voters who had selected a particular candidate reporting that another candidate had shown up on the screen. *(USA Today: 11-04-04)*

Matt Zimmermann, a lawyer working with the Electronic Frontier Foundation has reported receiving *a substantial number of voter complaints about touch screen machines recording the wrong vote.* He was of the opinion that this could be due to the fact that the voting machine screens were not properly calibrated. *(Culmers et. al.: Wall Street Journal: Nov 3, 2004)*

The Election Incident Reporting System (EIRS) *has received many reports from voters and election officials of votes for Kerry being recorded as votes for Bush.* (Wikipedia: 2004. Controversy: 3)

In Youngstown, it is reported that *electronic voting machines transferred an unknown number of votes from Kelley to Bush.* (Hill: 2006:18) An explanation as to why malfunctioning of voting machines happens has been offered by Alfie Charles of Sequoia Voting Systems, the company that manufactured the touch screen voting machines used in Pinellas, Palm Beach and two other Florida counties: *the machine's monitors may need to be recalibrated periodically....poll workers are trained to perform the recalibration whenever a voter says the touch screen isn't sensitive enough. It is said that the most likely reason for the screen showing the wrong candidate could be the fact that the voter had pushed the wrong part of the touch screen.* (USA Today: 11-04-04) This reflects badly on the certification authorities to have certified machines which are not simple to work on and which can easily malfunction.

This fault of recording what is not voted for is something which appears to be a fraud. It is interesting to note that no one has reported that Kerry appeared on the screen when they voted for Bush.

BREAK-DOWN OR FAILURE OF VOTING MACHINERY
& MISTAKES IN RECORDING VOTES, TABULATING AND
COUNTING.

In Cleveland, *a mistake in precinct poll coordination led to
hundreds of presidential votes being cast for a third party candidate
instead of the intended candidate.* (Fitrakis: 2005/01/03)

In Gahanna, Franklin County in Ohio some 20,000 were eligible
to vote and while the reported turn out was 70%, as much as
21,000 had voted according to the voting machines. *(Wikipedia:
2004....Controversy:12)*

In Broward County, Florida, a computer glitch, miscounted
thousands of absentee ballots just two days after the election.
(Chardy & Kidwell: Nov 27, 2004)

It is also reported that in a particular precinct, in the same
County *the unofficial results had Bush receiving 4,258 votes to
Democrat John Kerry's 260 votes. Records show only 638 voters cast
ballots in that precinct. Bush's total should have been recorded
as 365.* (AP: Nov 05, 2004) This County used Donaher Control
Inc's Electronic 1242 a relatively older style touch screen voting
system.

An instance is recorded where voting machines malfunctioned
when the removable cartridge which holds the tallies of the voting
machine was *loaded onto a laptop (and) transferred by secure
data lines to the County.* Matthew Damschroder, Director of the
Franklin County Board of Elections said that *the malfunction
occurred when one machine's cartridge was plugged into a
laptop computer and generated faulty numbers in several races.*
Kim Brace, the President of the consulting firm, Election Data
Services stated that *it is possible the fault lies with the software
that tallies the votes from individual cartridges rather than the
machines or the cartridges themselves.* It was found that this
type of cartridge was not used in other voting machines in Ohio.

(AP: Nov 05, 2004)

In Guilford County, *"vote totals for president were off by 22,000 votes."* (*North Carolina's Ballot Blues: 12-04-2004*)

The Electronic Frontier Foundation's Verified Voting Org, stated that *over 20% of the machines tested by observers around the country failed to record votes properly.* (*USA Today, 'E-voting ... Nov 3, 2004*)

One Roberta Harvey of Clearwater, Florida has reported that *she had tried at least half a dozen times to select Kerry and Edwards ... After 10 minutes trying to change her selection said she called a poll worker ... it took about ten attempts to select Kerry before and a summary screen confirmed her intended selection.* (*USA Today: 11-04-04*)

In Volusia County, Florida, *a memory card in an optical voting machine failed Monday at an early voting site and didn't count 13,000 ballots Officials planned to feed the ballots which voters fill in and count them Tuesday.* (*USA Today: E-Voting ... Nov 03, 2004*)

In Volusia County, a defective chip was found. *A computer chip is getting the blame for some voter problems in Volusia County. Those ballots will have to be re-fed. The defective chip was found Monday morning as poll workers fired up the machine for the last day of early voting .... The chip was escorted by Deputies to Daytona Beach and is in use right now.* (*The Command Post: Nov 18, 2004:3*)

ABC13 reported that electronic voting machines had broken down in Harris County in Texas, *creating long lines and causing confusion for voters and poll workers. The problems were eventually fixed but some voters left due to the delay.* (*ABC13EveningNews: www.demos-usa*)

In Broward County, authorities discovered that one electronic voting machine started subtracting votes after reaching 32,500 votes. (Chardy & Kidwell: Nov 27, 2004). It is interesting to note that the manufacturer E&S has admitted that they *have known about (but not rectified) this issue for two years since the same problem had arisen in a previous Mayoral Election. (BrowardCo.,FL) (Wikipedia: 2004 ... Controversy:3)*

In Broward County, Florida, 21 of its 6020 touchscreen electronic voting machines were found to be out of order and had to be removed. *(SanJoseMercuryNews: Nov 03, 2004)* The question arises as to how such faulty machines had come to be certified.

In Carteret County, North Carolina *a voting machine used to store electronic ballots ran out of storage space and 4,438 votes 'disappeared'. Because the electronic voting machines used in Carteret County do not count or create a paper ballot, the disappeared votes were irretrievably lost. (Bollyn:Conspiracy Planet: Dec 5, 2004)*

In Perry County, *a punchcard system reported about 75 more votes than there are voters in one precinct. Workers tried to cancel the count when the tabulator broke down midway through, but the machine instead double counted an unknown number in the first batch. (AP: Nov 05, 2004)*

Voting machines in South Carolina *broke down and had to be replaced by paper ballots (USAToday: www.demos-usa.)*

In South Carolina, in a handful of precincts, *voters were forced to switch to paper ballots while technicians got the Votrionic touch screens from ES & S up and running within about 90 minutes. (USA Today: E-voting : Nov 03, 04)*

What is also important is how certification standards were ensured after the technical adjustment of the software.

In Snohomish there were over 50 complaints re *touch screen calibration ... the screens were off calibration on some areas of the ballot but not on others. This is true of most of the Sequoia touch screen complaints ... for sometime Snohomish County knowingly used uncertified firmware. (Crew: Dec 3, 2004)*

The Election Protection Coalition on 13th November 2004, held a Hearing at the New Faith Baptist Church in Columbus, Ohio *where for three hours burdened voters, one after another, offered sworn testimony about election day voter suppression and irregularities that they believe are threatening democracy ... Many of the voters testified were clearly Democrats who wonder if their losing presidential candidate Senator John Kerry was able to draw all the votes that were intended for him. (Wikipedia: 44, from http://cleveland.com/news/plaindealer/index.ssf?/base/news/110042844428640.xml)*

In Gaston County, North Carolina it was found that in making the totals the voting machines *omitted an entire precinct of 1209 votes(North Carolina's Ballot Blues, 12-04-04)*

In Gaston County, when an alarming 12,000 votes were lost without any trace, *the Election Director hired a voting machine technician to upload the county vote totals and did not oversee the process. (North Carolina's Ballot Blues, 12-04-04)* The certification process appears to have been ignored.

In Miami County, Ohio the explanations asked from the Ohio Secretary of State included a query as to how 19,000 votes were added to election totals which had been missed by the counts of the voting machines. *(House Judiciary Committee: Dec 4, 2004(1))*

A Diebold contractor at the Georgia Warehouse had said that 25 to 30% of the machines were either crashing as they were being booted up or otherwise failing. *(Gumbel: 2005:236)*

In Sharpy County, Nebraska, a single voting machine recorded 10,000 extra votes. *(Bollyn: Conspiracy.: Dec 5, 2004)*

In Utah, a programming glitch in the punch card counter dropped 33,000 ballots from the totals. When a recount was done days later the 33,000 votes were distributed to the candidates to whom they were cast. In the words of Kristen Swensen, the Utah County Elections Coordinator, *it was just the way it was programmed initially. (The Command Post: Nov 18, 2004:3)*

In LaPorte, Indiana, in the counting of ballots *it was noticed that the first two or three printouts from individual precinct reports all listed an identical number of voters. Each was listed as having 300 registered voters. That means the total number of voters for the county was 22,200 although there are actually more than 79,000 registered voters ... the patch from ES&S didn't work-they might have to manually input the information. (The Command Post: Nov 06, 2004:16)*

In Santa Clara County, California, *some cards used to activate the machines got stuck temporarily and some machines could not be used because they were matched with the wrong activator cards. (SanJose MercuryNews: Nov 03, 2004)*

Election officers in Raleigh, North Carolina discovered in October 2004, that *early voters had to make several attempts to record their votes on ES&S Systems. Officials compared the number of votes counted and realized that 294 votes had been lost. (Infernal Press.Com 11/29/2004)*

Broken down electronic voting machines were reported in Iowa (Des Moines Register: www.demos-usa.org/page 196.cfm:1)

Voting machines breaking down has been reported at New York (NewYork Sun: www.demos-usa.org/page196.cfm)

It is reported that in many areas around New Orleans, there were many instances of broken voting machines and *at one precinct all three voting machines are broken and voters have been told to come back later. (Nola.com: www.demos-usa)*

Teed Rockwell states that in the presidential election, *29 precincts in Cuyahoga County, Ohio reported votes cast in excess of the number of registered voters—at least 93,136 extra votes total and the numbers are right there on the Cuyahoga County Board of*

Elections Web Site. (Rockwell: Conspiracy.: Dec 5, 2004)

In Ohio, it is reported that 92,000 ballots failed to record a vote for the President on punch card machines. *(Hill: 2006:18)*

Sandusky County Elections Director, Barb Tuckerman believes the votes were counted twice when they were mistakenly placed alongside a pile of uncounted ballots. The room where the ballots were being fed into optical scan machines on election night was so crowded that ballots had to be placed on the floor, Tuckerman said ... The problem was discovered when Tuckerman had found that one precinct showed 131% of registered voters had cast ballots. *(The Command Post: Nov 18, 2004:3)* This is also confirmed by Jeanne Cummings. *(Cummings: Wall St. Journal: Nov 19, 2004)*

Evidently proper care has not been taken of ballot papers. In my own experience, it is a ludicrous situation for ballots to be placed on the floor. This should never have happened.

News Analysts from The Washington Despatch reported that in the Palm Beach County the official website posted that 542,835 votes were cast at the election. However, only a total of 454,427 voters had turned out at the election and this number included the absentee ballots that had been cast. This leaves a vast discrepancy of 88,408 votes cast at the presidential election. *(NewsAnalysts: WashingtonPost: Nov 05 ,2005)*

Sherole Eaton, the Deputy Elections Director in Hocking County, Ohio states that 3 days before the election an employee of the Triad Co, the company that provides voting machines to 41 Ohio counties including Ohio saw her in her office for the purpose of preparing voting machines for the recount and wanted the names of the counties where recounts would be done. This was contrary to a statutory requirement which specified that counties should be selected at random. This officer gained access to the machines and repaired or reprogrammed the main computer.

Eaton protested and even Congressman John Conyers requested an FBI investigation into this incident. However Eaton ended in hot water, backed off, and was subsequently fired from her job. *(Nolan: 2004)* The fate that befell Eaton, can be contrasted with the action taken against county election supervisors at the Seminole and Martin Counties in the 2000 Presidential Election where it was proved that they had allowed thousands of absentee ballots to be handled and altered by Republican officials-where although there was complete proof of complicity to boost the votes of a particular political party, no action was taken against the officials.

In Miami County, Ohio, an officer who had functioned directly in charge of the election reported that the *electronic vote counting system included votes never cast in the total vote count reported for the 2004 Presidential Election. (Peckarsky et al.: Nov 6, 2006)* The machines used were optical scanner machines.

On the whole it can be said that the evidence of malfunctioning voting machines justifies the statement by Keith Jennings, Director of Count Every Vote 2004:

> *While the U.S. of America is a strong democracy it is also a flawed democracy. (Gross: Nov 6, 2004)*

What is important in the very high incidence of malfunction in the electronic voting machines is the extent to which an election result can be varied. Well over 80% of the ballots were tabulated and totaled by electronic voting machines and this is indicative of the type of result that could be expected.

3.6

SABOTAGE EVIDENCE RE ELECTRONIC VOTING MACHINERY

THERE IS EVIDENCE POINTING to the fact that certain malfunctioning had been done premeditatedly, which amounts to sabotaging the outcome of the election.

In New Jersey, in the City of Passaic, 75% of electronic voting machines failed to work on election day. A Philadelphia Voting Machine Supervisor V. Thomas Mattia who had examined the machines had said that this was due to sabotage which made the Democratic candidate call the FBI to investigate. But strangely later the Supervisor had backed out with no reason. *(BlackBoxVoting: 30)*

Black Box Voting. Org stated that in King County, Washington, the primary election, six weeks ago, *uncovered an internal audit log containing a three hour delection on election night, 'trouble slips' revealing suspicious modem activity and profound problems with security, including accidental disclosure of critically sensitive remote access information to poll workers, office personnel and even in a shocking blunder to Black Box Voting activists. (USA Today:' E-voting: Nov 3, 2004)*

It is reported that *the Bush supporters who ran the Central Polling Station in Ohio's Warren County forced out the Press and poll monitors so they could count the vote in secret (The Crisis Papers).*

As pointed out earlier the counting of ballots is a public task and there should be total transparency–the counting should be done in the presence of the candidates. There cannot be any counts done with secrecy. Secrecy in the counting of ballots has to be mandatorily banned by law.

Lynn Landes, one of America's leading journalists on voting technology and democracy issues states of how the Republicans have weighted in favor of electronic voting systems without any consideration for their malfunctioning. This can even amount to sabotaging the democratic right of people to have their voice when one realizes that when the U.S. Commission on Civil Rights met on April 9th 2004 to examine the "integrity, security and accessibility in the Nation's readiness to Vote", the three Republican commissioners walked out quoting personal reasons. Landes comments:

It appears that voting technology is a topic that the Republican leadership wants to tightly control.

It is well known that Republicans own and control most of the companies that manufacture electronic voting machines. It is even said that *The Bush Administration has stacked the Election Assistance Commission with supporters of paperless voting technology, while the National Institute of Standards and Technology (NIST) got walloped with a $ 22 million budget cut in fiscal 2004, which means that NIST will have to cut back substantially on its cyber security work as well as completely stop all work on voting technology for the Help America Vote Act. (Landes: Lynn Investigates: 4/13/2004)*

It is alleged that the Republicans were behind the malfunctioning of voting machinery and that the elections were rigged through the manufacturers. This is quite possible when one considers the numerous instances when the electronic voting machines have malfunctioned as has been documented in this book and by many other critics. It is well known that technicians can program for the tabulation of totals to vary from the actual totals. It is a fact that the certification of electronic voting machines was a farce in that technicians had to be called in whenever the voting machines malfunctioned and these technicians had to work on their own without any security supervision by competent computer experts, who could alone be certain that the codes had not been altered to falsify the outcome. Ensuring standards in certification have been totally ignored in the attempt to get the machines working as soon as possible. In this process certification has been compromised.

The problem lies in the very high incidence of malfunctioning and the nature of the discrepancy–at times going into not hundreds, but tens of thousands of votes–numbers that can easily bring about a victory.

While in the case of electronic voting machines the method used appears to be malfunctioning through the software, the situation was different in the case of punch card machines. Here, when overused, due to wastage in use, the metal pins that pushes out the chad, can no longer push out the chad. This results in a ballot without a vote for the President. It is alleged that, *These (African American) precincts were allocated a disproportionately high proportion of punch card voting machines compared to other precincts. High turnout and a very high percentage of the voters voting the same (Democrat) resulted in an anonymously high concentration of reported broken punch card machines, machines in which the metal pin that punches out the chad can no longer push the chad through the hole because too many chads had built up beneath it–resulting in ballots without a vote for President.* (*Wikipedia: 2004 US Presidential*)

In the 2004 Presidential election in Ohio more than 90,000 votes were discounted due to hanging chads. (*Coalition Against*

Election Fraud: 2005-01-06) Over 92,670 ballots in Ohio registered no presidential vote, *(WSWebSite, Editorial: Nov 24, 2004)*. This is also confirmed by another journalist, Hunter who states that there were 92,672 ballots on which no vote for President was recorded. *(Hunter: Nov 05, 2004)*

This is a matter which should have been addressed by the Elections Departments of every State in order to avoid a repetition of what happened in the 2000 Presidential Election. In the case of the 2000 Presidential Election it was found that on punch card machines there were a high percentage of ballots that did not record a vote for the President. *(Karunaratne: 2004: 74–76.)* In fact, The Wall Street Journal hinted sabotage when they said; *Chad build up must be a special problem with presidential choices.* (Dec 04, 2000). The fact was that the punch card machines had punched clean for Congress and Senate elections.

The non replacement of worn out parts can reflect wanton and deliberate action. In my Chevrolet Roadtrek when oil has to be changed, it is indicated on the dash board. In my Honda Accord even a burnt out bulb is indicated on the dash board. Similar devices and alarms should be built into the electronic voting machines. This is a basic necessity that has been ignored. This reflects the inefficiency of the Elections Administration.

It is important to note that sabotage can easily take place when important tasks in holding public elections are handed over to outside bodies and no proper control or supervision is enforced. In the words of Lynn Landes: *State Election officials across the country have outsourced the tabulation of the vote to a handful of Republican and foreign owned corporations. There is no meaningful public oversight of the count. No one knows if votes are being added, subtracted or switched. (Landes: Did Networks Fake. 2005)*

This Study has repeatedly pointed out that the process of holding elections is in the hands of technicians in the pay of the manufacturers of voting machines and no supervision is being imposed on their work by the authorities that should be certifying the machines. It is important to note that the manufacturers of the voting machines are supporters of the Republican Party.

REVELATIONS OF THE INADEQUACY OF
VOTING MACHINES

THE INADEQUACY AND UNRELIABLE nature of electronic voting machines has come up again and again.

The authorities in California have been dissatisfied with the electronic voting machines supplied for the 2004 US Presidential Election. James Dunn, a former Diebold technical worker had said that the company was aware of technical problems in the machines but ignored them. In his words: *The machine would lock up or lose its software load. A very uncommon thing and not a good thing. And once that machine is locked up you're unable to produce voter cards which means you are unable to open the election voting machine and people can't vote but they shipped it anyway. (Wikipedia: 2004. Controversy: 41)*

On October 27, 2004 it was decided that 15,000 brand new touch screen voting machines should not be used. Election officials stated that there were serious flaws with the machines and that Diebold repeatedly misled the State about them. The California Secretary of State said that *(Diebold) literally engaged in absolute deplorable behaviour and to that extent put the election at risk, jeopardizing the outcome of the election. (Wikipedia: 2004. Controversy: 41)*

The California Attorney General stated that he will sue e-voting technology maker Diebold on charges that it defrauded the State with aggressive marketing and overstated claims and sold the State poor quality equipment that did not produce a paper trail and was full of security vulnerabilities. *(Wikipedia 2004. Controversy: 42)*

In *Analysis of an Electronic Voting System*, Tadayoshi Kohno of the Department of Computer Science at the University of California at San Diego, Adam Stubblefield and, Aviel D. Rubin of The Information Security Institute at John Hopkins University and Dan S. Wallach of Department January 2003, of Computer Science of Rice University, state that: *this voting system is far below even the most minimal security standards applicable in other contexts.* (Kohno et al.: 2004:3) It is important to note that they even add that as far as the AccuVote terminals are concerned, *voters can easily program their own smartcards to simulate the behaviour of valid smartcards used in the election. With such homebrew cards a voter can cast multiple ballots without leaving any trace. (Kohno et al.: 2004:4)*

In their conclusive report they conclude:

We found significant security flaws; voters can trivially cast multiple ballots with no built in traceability, administrative functions can be performed by regular voters and the threats posed by insiders such as polls workers, software developers and janitors is even greater. (Kohno et. al: 2004:21)

Researchers at John Hopkins and Rice Universities have released a Report that reveals *numerous programming faults and security vulnerabilities in the source code for Diebold's Accu Vote-TS Voting Machines. (Zetter: 08-12-03)*

The *Washington Times* reports how when the State of Maryland employed a computer expert to look into the security of voting machines he had *broken into the computer at the State*

Board of Elections during a test and completely changed the election results. (Lott Jr: Washington Times: May 10, 2004)

Goldfarb documents how a fictional gubernatorial race conducted revealed that *it would take only one person with a sophisticated technical knowledge ... to change the outcome........ the three major electronic voting systems in use have significant security and reliability vulnerabilities. But it added that most of these vulnerabilities can be overcome by auditing printed voting records to spot irregularities. (Goldfarb: WashingtonPost.Com: june 28, 2006)*

When one studies the complicated nature of how the voting machines have been programmed, it is no cause for surprise that irregularities have cropped up. The Analysis of the electronic voting system by specialists, Kohno et. al. tells us in no uncertain terms of the complicated nature of the system:

To get started the voter must have a voter card. The votercard is a memorycard or smartcard i.e. it is a credit card sized plastic card with a computer chip on it that can store data and in the case of the smartcard, perform computation. ....The voter takes the votercard and inserts in into a smartcard reader attached to the voting terminal. The terminal checks that the smartcard in its reader is a votercard and if it is, presents a ballot to the voter on the terminal screen. The actual ballot the voter sees may depend on the voter's political party which is encoded on the voter card ....At this point the voter interacts with the voting terminal touching the appropriate boxes on the screen for his or her desired candidate.... Before the ballots are committed to storage in the terminal the voter is given a final chance to review his or her selections. If the voter confirms this the vote is recorded on the voting terminal and the voter card is cancelled. The latter step is intended to prevent the voter from voting again with the same card.

They add that an adversary *could program a smartcard to ignore the voting terminal's deactivation command ... could use one card to vote multiple times (Kohno: 2004: 10)*

Regarding the security codes used they state that there is little difference in the way the code is developed for voting machines relative to other commercial endeavours. (Ibid: 21)

Kohno et. al. state that when the State of Maryland got SAIC and Raba, two computer companies to perform independent analyses of Diebold's AccuVote System it was found that the smartcards had been modified *so that a voter could vote more than once. (Kohno et al.: 2004:5)* This action actually reduces the certification process to utter ridicule. This could perhaps explain how in some instances the number of ballots were more than the number of voters that had signed in.

The impact that faulty machines can have on an election is revealed by the extent to which electronic voting machines that are likely to malfunction have been used at the election. Maryland purchased 5000 Diebold Touch Screen Terminals at a cost of $ 12 million in March 2002. Maryland also gave a further contract of $ 55.6 million *to provide and service 11,000 additional Diebold machines to be used throughout the State for next Spring's Presidential Primary. (Zetter: Aug 12, 03)*

Electronic Voting Machines store the information, i.e. the markings on the ballots electronically. This is generally known as Direct Recording Election (DRE) Software codes are used by the manufacturers and they claim that the software codes are theirs and cannot even be inspected. In view of the extensive evidence re malfunctioning of voting machines it is extremely necessary that the codes have to be checked in the certification process and further action has to be taken by the certifying authorities to introduce codes designed to ensure that the certified codes cannot be altered.

Barbara Simmons, Past President, Association for Computing Machinery is quoted:

> *Anyone with access to the electronic software of a major voting machine vendor can change the outcome of a national election and determine which party will control Congress; Election Fraud can now be committed on a national, not just a local basis.* (The Crisis Papers)

Douglas Kellner, a New York Elections Commissioner has the following scathing remark against the coding systems in the electronic voting machines:

> *Using electronic voting machines to count ballots is akin to taking all the paper ballots and handing them over to a couple of computer tech people to count them in a secret room and then tell us how it came out. This is not an acceptable way of conducting elections in a democracy"* (Quoted by Lefebvre: Dec 24, 2003)

In January 2003, Bev Harris, an owner of a small Seattle based public relations firm while researching the web stumbled into *an open file transfer protocol site containing the source code for Dieold's Accu-VoteTS as well as program files for the company's election management software.* Harris found that *it was possible to enter the voter database using a standard Microsoft application, Microsoft Access and change votes without leaving any trace.* (Gumbel: 2005:252) With this finding Avi Rubin of John Hopkins University was able to hack into the system; they found that the security systems used were so poor—*the password unlocking the system's encrypted data was written directly into the source code. This was a violation of the most rudimentary principles of cryptography. (Gumbel: 2005:253).* In other words with the access to the source code any Diebold voting machine could be interfered with.

Professor Avi Rubin of John Hopkins University who analysed Diebold's 47.609 lines of code found that it used an encryption key that was hacked in 1997 and is no longer used in secure computer programs. *(Wikipedia: 2004. Controversy:7)*

Avi Rubin states:

I have been saying all along that my biggest fear is that someone would program a machine to give a wrong answer. If that were to happen the machine would still work fine—we just would not know it. (AP: Nov 04, 2004)

David Jefferson a computer scientist who advised Kevin Shelley, the California Secretary of State said that *Having a vote misrecorded by a computerized voting machine is not a far fetched concern at all (San Jose Mercury News: Nov 03, 2004)*

Ken Clark, Diebold's Principal Engineer in October 2001, stated that the audit system of Diebold machines *could be accessed from the outside– a blatant security breach....can be accessed without a password and changed using Microsoft Access. (Gumbel: 2005:258)*

Ion Sancho, the Elections Supervisor in Leon County, Florida had obtained the services of a Finnish computer expert and the tests showed that *the election workers could alter the vote tallies by manipulating the removable memory cards in the voting machines and do so without detection. (Whoriskey: March 26, 2006)*

A hack test done officially on Diebold machines *changed the results of an election from 2-6 to 7-1, left no traces of evidence behind.* On seeing this, Ion Sancho had decided that Leon County will not use Diebold voting machinery thenceforth. *(Brad Friedman: Dec 18, 2005)*

Maryland's State Board of Elections commissioned two studies on the software used in the Diebold machines. The first study done by a computer risk assessment company, SAIC International, found as much as 328 security weaknesses, 26 of them critical. The second study done by Raba Technology, a security company that employs many former National Security Agency staff, found that *it took approximately twenty seconds to pick the two locks securing each of Maryland's sixteen thousand AccuVote-TS terminals and every one of the locks–thirty two thousand in all– was identical.* In the words of William Arbaugh of the University of Maryland, *We could change the ballots (before the election) or change the votes during the election.* (Gumbel: 2005:260)

The manipulation of electronic voting machines can easily take place when one considers that proper security has not been ensured. Princeton University computer scientists have found that *Diebold uses the same key to open all machines.* In fact it is found that the Diebold website carries a photograph of the key. (Brad Friedman: Jan 24, 2007) In the words of David Jefferson, a security expert at the Lawrence Livermore National Laboratory, and a consultant to the California Secretary of State:

> *What (Diebold) did is create a big complex building, put locks on every door, use the same key for every lock and then publish a picture of the key on the wall.* (Quoted by Gumbel: 2005:255)

In Riverside County California, at a test, Dr Michael Shamos, of Carnegie Mellon University when testing Sequois voting machines was able *"in an instant to transform a handful of votes into thousands"* (Friedman: Dec 6, 2006)

The Report by The information Security Institute at John Hopkins University has concluded against the codes used in the Diebold machines as follows:

The AccuVote-TS machines would allow a voter to cast multiple votes and was vulnerable to someone hacking into the system to switch votes. The researchers also found that cryptography wasn't written into the code in some places where it should have been used and where it was written into the code it was used poorly and incorrectly. (Zetter: Aug. 12, 2003)

Professor Rubin one of the authors of the Report had said that the code that has been used *is so full of mistakes and misunderstandings and improper use of cryptography that it was obvious to us that the person who wrote this code had no training.* Diebold replied to the effect that the code has been revised to which Professor Rubin has doubted whether the code could have been corrected that soon. In his words:

I do not think anybody has the capability to develop a whole new system from scratch in a year. And I don't think Diebold had any incentive to do so because none of this news broke until recently. The only alternative is that they fixed and I don't think it was fixable. (from Zetter: Aug 12, 03)

It is interesting to note that the security systems used in electronic voting machines in the USA have been found wanting compared with security systems used in other countries. Press comments by International Monitors in Florida included comments: *voting procedures fell short in many ways of the best global practices.... They had less access to polls than in Kazakhstan, that the electronic voting had fewer fail-safes than in Venezuela, that the ballots were not so simple as in the Republic of Georgia and no other country had such a complex national election system.*

The International Monitors comprised two members fanned out in 11 States and included citizens from 36 countries ranging from Canada, Switzerland, to Latvia, Kyrgyzstan, Slovenia and Belarus.

The sum total of irregularities documented can be summed up in the words of USA Today:

> *Nearly one in three voters including about half of those in Florida were expected to cast ballots using ATM style voting machines that computer scientists have criticized for their potential software glitches, hacking and malfunctioning. ('E-Voting Irregularities'. USA Today: Nov 3, 2004)*

Malfunctioning electronic voting machines have been very common. In this situation one has to agree with Bev Harris's statement that with computerized voting, the certified and sworn officials step aside and let technicians and sometimes the county computer guy tell us the election result. As detailed by me many are the instances of malfunction in electronic voting machines and many are the instances where chips were flown in and action taken to correct the software of the computer. In such instances the entire certification process is reduced to utter ridicule.

In the case of Diebold machines, it was well known to Diebold technical staff that the totals could *be modified remotely via undocumented backdoor in central tabulator. (Friedman: Sept 15, 2005)*

I by my experience in handling the election process in Sri Lanka as an officer of the Sri Lanka Administrative Service handling elections as the officer in charge of a number of precincts–about two dozen precincts and having functioned in charge of the counting of ballots in two electorates in my capacity as an Assistant

Returning Officer I can speak with authority. Every ballot paper has to be checked to ensure that the vote cast was appropriately tabulated according to the wish of the voter. The counting has to be done very systematically and meticulously. It is the height of absurdity to enable technicians and computer specialists to have a hand in providing off-hand solutions and computer repairs to get the results. It is a properly supervised and securely done hand count that can assure accuracy in the tallying of the vote. I can make this statement without any reservation.

3.8

INSUFFICIENT VOTING MACHINES

W HILE IT IS A difficult task to ensure that all electronic voting machines used do not malfunction in use, it is a relatively simple task to find out the number of voting machines that are required to enable the voters to cast their ballots. The number of possible voters has to be assessed; the number of voters who could cast their ballot in an hour, multiplied by the number of hours of polling would indicate the total number of voters that could be serviced with one voting machine. An addition of a few machines has to be made for use in case of a break down. This is a simple arithmetical calculation. It is sad that the Administration has failed even in this fundamental basic task in the case of the 2004 Presidential Election.

In the 2004 US Presidential Election, in Franklin County of Ohio, a Study by the Democrats of the House Judiciary Committee found that 125 voting machines that were supposed to be available according to records were not available on election day. This was in the City of Columbus and caused delays of two to seven hours. In certain precincts the number of machines available were less than in the primaries where there would have been far less voters. At Kenyon College in Knox County, just outside Columbus, there were only two voting machines to service one thousand three hundred voters and that meant that if everyone turned up the voting time had to take thirty hours. The voting had continued till 2.30 A.M. the next day. (Gumbel: 2005: 291) In Sri Lankan parliamentary elections in the days when I was in charge, before 1973, a very senior officer of the rank of Assistant Returning Officer of the District had to personally do a physical check on the availability of the required strength of staff and the ballots etc on the evening of the earlier day when there was ample time to rectify matters. The Assistant Returning Officer had wide powers to draft support staff and order anything that had to be done to rectify matters.

In Columbus, Ohio where voters had to wait long in lines, at a Public Hearing on Election Irregularities and Voter Suppression, held by non partisan voter rights organizations, on November 13, the documents produced proved that as much as 68 electronic voting machines were in storage and were not used on election day. It was reported that voters had to wait 2 to 7 hours to vote. What is more important is that:

> An Analysis of the Franklin County Board of Elections' allocation of machines reveals a consistent pattern of providing fewer machines to the Democratic City of Columbus, with its democratic mayor and uniformly democratic city council despite increased voter registration in the city. The result was an obvious disparity in machine allocations compared to the primarily Republican white affluent suburbs.

In fact it is stated that in the Democratic stronghold of Colombus, 179 of the 472 precincts had at least one and up to five fewer machines than in the 2000 Presidential Election. (Wikipedia: 2004. Controversy:37) This is a serious allegation for which the Elections Department at the State level should be held responsible for not supervising the allocation of voting machines.

In Ohio, when confronted with the fact that many precincts did not have sufficient voting machines, Kenneth Blackwell the Secretary of State is reported to have replied at a news conference that it was the responsibility of the Counties to ensure that there were sufficient voting machines. (Gumbel: 2005: 292) It has to be mentioned that the Secretary of State cannot absolve himself of responsibility as he is the officer in charge of the election in the entire State. The county supervisors of elections though elected by the people come under his direct supervision; their actions have to be monitored carefully by the Elections Administration, headed by the Commissioner for Elections at the state level directly coming under the Secretary of State and appropriate remedial action taken. It is for this supervisory task that the Secretary has an Elections Department at State level, under his direction.

The executive summary of the Democratic National Committee Report states:

> *Scarcity of voting machines caused long lines that deterred many people from voting. Three percent of voters who went to the polls left their polling places and did not return due to the long lines. (Kennedy Jr: June 6, 2006)*

It has to be mentioned that the denial of voting machines in sufficient numbers can be construed to be sabotage deliberately done.

Isn't it sad that the very high incidence of malfunction in electronic voting machinery and manipulation in the allocation of machinery to sabotage true voting makes one agree with Rev. Jessie Jackson:

> We can live with winning and losing. We cannot live with fraud and stealing.... The integrity of our election process is on trial.

4

THE EXIT POLLS

EXIT POLLS ARE POLLS that are carried out by organizations with the intention of providing an indication of the results of an election. The system of conducting Exit Polls has been developed to be an art in itself. This is done by people who question the persons that have just completed voting and analyze the results. Though this may not perfectly indicate the true results, it has been generally found that Exit Polls have been indicative of the true results. Professor Steve Freeman and Josh Mittledorf state that *the reliability of Exit Polls is so generally accepted that the Bush Administration helped pay for them during recent elections in Georgia, Belarus and Ukraine. (Freeman: 2005)*

In many instances Exit Polls have been found to predict the actual result. In the 2003 Salt Lake County Mayoral Race

The results are very precise.... The KBUY/Utah Colleges Exit Poll predicted 53.8% of the vote for Rocky Anderson and 46.2% for Frank Pignanelli. In the actual vote Anderson carried 54% to Pignanelli's 46% (Wikipedia: 2004. Controversy: 31)

In the 2004 US Presidential Election, the Exit Poll on Election Day at 12.23 A.M. on the results done by Edison Media Research and Mitofsky International for the National Election Pool, predicted Kerry as the winner of the popular vote by 5 million. However the official actual result gave George W. Bush a victory by 3 million. There was an 8 million vote difference between the exit polls and the actual results.

According to the Exit Polls, it was said that Kerry was having a very comfortable lead in six States.

Further, of the Exit Polls, one can quote a Exit Poll Analyst-Jonathan Simon for the methods used:

"his methodology was, as the night wore on, to mix an actual tabulation data with the initial exit poll data in such a way that by the time the full count was in, the 'exit poll' would conform very closely to the 'actual' vote (Internet correspondence, Nov 6, 2004) He notes that the data may have already been adjusted to match counts, but were probably still pure. If they already had been adjusted, it means that the pure poll numbers favoured Kerry to an even greater extent." (Freeman: 04/11. Paper 4296, also Wikipedia: 2004. Controversy: 33)

An Analysis of the Exit Polls and the actual vote in the battleground states indicate clearly how there is a discrepancy between the Exit Polls and the actual vote.

It is observed that in the State of Colorado the Exit Poll predicted differential of 1.8% in favor of Bush turned out to be a 5.2% differential in favor of Bush in the final tally. In Florida The Exit Poll differential of 0.1% in favor of Bush turned out to be a differential of 5.0% in favor of Bush. In Iowa, the Exit Poll differential of 1.3% in favor of Kerry turned out to be a differential of 0.9% in favor of Bush in the final tally. In Michigan the Exit Poll differential of 5.0% in favor of Kerry turned out to be 3.4% in favor of Kerry in the final tally. In Minnesota the Exit Poll differential of 9.0% in favor of Kerry turned out to be a 3.5% differential in favor of Kerry. In Nevada, the 1.3% lead that Kerry had in the Exit Poll turned out to be a lead of 2.6% in favor of Bush in the final tally. In New Hampshire the 10.8% lead that Kerry had in the Exit Polls turned out to be a lead of only 1.3% for Kerry. In New Mexico the 2.6% lead that Kerry had in the Exit Poll turned out to be a 1.1% lead for Bush. In Ohio the 4.2% lead that Kerry had in the Exit Poll turned out to be a 2.5% differential in favor of Bush in the final tally. In Pennsylvania the 8.7% lead that Kerry had in the Exit Poll turned out to be a lead of only 2.2% in the final tally. In Wisconsin, the 0.4% lead that Kerry had in the Exit Poll remained constant in the final tally. (*Wikipedia: 2004. Controversy: 34*)

On the whole it is seen that while the Exit Polls had been definitely in favor of Kerry, in the actual tally the votes have been largely found in favor of Bush.

The exit polls in the case of the two major States of Florida and Ohio indicate that there were wide disparities between the results of exit polls and the actual poll in the 2004 Presidential Election which cannot be attributed to be mere incidental. The actual polling in the various battle ground states for Kerry is below the predicted poll at the exit polls. Of the actual poll in the State of Ohio, it is said that, *Given that the Exit Poll indicated Kerry received 52.1% of the vote we are 95% sure that the true percentage he received was between 49.8% and 54.4%...... we are 99.5% sure that the true percentage Kerry received was at least 49.2% It turns out that the likelihood that he would have received only 48.5% of the vote is less than one in one thousand. (.0008)* (*Wikipedia: 2004. Controversy: 34*)

In respect of the poll in Florida, *the likelihood of Kerry receiving only 47.1%, given that the exit polls indicated 49.7% is less than three in one thousand. (.0028) (Ibid)*

Dick Morris, a Republican pollster is of the opinion that in respect of the exit polls in the six battle ground States of Florida, Ohio, New Mexico, Colorado, Nevada and Iowa States: *Exit polls are almost never wrong.... To screw up one Exit Poll is unheard of, To miss six of them is incredible.* (Wickipedia: 2004 US Presidential Election Controversy: Exit Polls: 30)

Steven F. Freeman states that *the deviations between exit poll predictions and vote tallies in the three critical battleground states could not have occurred strictly by chance or error and that no solid explanation has yet been provided to explain the discrepancy.* (Wikepedia: 2004. Controversy: 35)

On the whole the evidence regarding the variation between the results of Exit Polls and the actual results coffered up by the electronic voting machines fairly confirms the opinion of Jonathan D. Simon: *there is little to suggest significant flaws in the design or administration of the official exit polls.....an honest and fair voting process would have been more likely than not– at least 95% likely to have determined John Kerry to be the national popular vote winner of Election 2004.* (Simon: 2004)

Researchers at the University of California, Berkeley went through the results of the presidential elections 1996, 2000 and 2004 and concluded:

In the 15 counties (in Florida) using touch screen voting systems the number of votes granted to Bush far exceeded the number of votes Bush should have received in counties using—given all of the other variables– while the number of votes that Bush received in counties using other types of voting equipment lined up perfectly with what the variables would have predicted for those counties. The total number of excessive votes ranged between 130,000 and 260,000, depending on what type of problem caused the excess votes. The counties mostly affected by the anomaly were heavily democratic.

(Wikipedia: 2004. Controversy:35) http://ucdata.berkeley.ed)

They conclude that

The data show with 99.0% certainty that a county's use of electronic voting is associated with a disproportionate increase in votes for President Bush. (Wickipedia: 2004. Controversy: 36)

Will Pitt says: *The Exit Polls were dismissed as being inaccurate. But their accuracy may have depended on the type of voting machine that was used, implying that the machines, not the polls were inaccurate.*

Pitt's findings on Florida are that *the counties that used optical scanner machines to record votes showed a consistent pattern of far more votes for Bush and far less votes for Kerry than projected based on the number of registered republicans and democrats.... The counties that used touch screen voting had a relatively normal distribution of final votes for Bush and Kerry compared to the amount of registered democrats and republicans and the projected turn out. (Pitthttp://www.ideamouth.com/voterfraud.htm)*

USCountVotes.org states after a detailed comparison of the actual votes and expected votes per political party:

EXPECTED votes would normally vary from the ACTUAL votes due to increased voter turn out by one party, Independents voting REP or DEM or other factors. What seems very odd in these numbers is that the increase in ACTUAL votes from EXPECTED votes has a striking pattern of being so much higher for REPs than for DEMs in counties using optical scan machines, even when smaller counties are excluded from the analysis. (USCountVote4s.org: November 24, 2004)

This pattern of higher actual votes for Republicans in precincts where optical scanning machines were used finds collaboration in an analysis by Elizabeth Liddle. She states that *machine type was a significant predictor of percent changes in voting ... Counties using E-touch machines showed significantly positive percentage changes in vote for both Republican and Democrat candidates, with greater mean percentages for the Democrat. However Counties using Op-Scan machines showed significant positive change only for the Republican candidate, the mean change for the Democrat being insignificantly greater than zero. (Liddle:http://ustogether/org/elections04/Liddle.Analysis. html, 11-24-04)*

There has been evidence to the effect that the outcome of the vote count could not be normally expected.

In the Florida–Counties of Suwanee, Lafayette and Union, registered democrats outnumber Republicans 3 to 1 but in these counties Bush won heavily. How?

One internet site commented *"George W. Bush's vote tallies especially in the key State of Florida are so statistically stunning that they border on the unbelievable. (http://www.avrubin.com/vote .pdf)*

Thom Hartman, in 'Evidence Mounts that the Vote may have been hacked' states that in the areas where touch screen voting machines were used there was a close co-relation between the number of Democrat and Republican voters and the actual vote for Kerry and Bush, in the areas where optically scanned paper ballots fed into a central tabulator personal computer, there was no co-relation. Hartman quotes the following anomalies:

1. In Baker County where 69.3% were registered Democrats and only 24.3% were registered Republicans the actual vote was only 2,180 for Kerry and 7,738 for Bush.
2. In Dixie County where of 4,988 voters 77.5% were registered Democrats and only 15% were registered Republicans, in the actual vote only 1,959 voted for Kerry while 4433 voted for Bush.

Hartman states:

The pattern repeats over and over again but only in the counties where optical scanners were used. Franklin County-77.3% registered Democrats went 58.5% for Bush. Holmes County-72.7% registered went 77.25% for Bush. Yet in the touch screen counties where investigators may have been more vigorously looking for such anomalies, high percentage of registered Democrats generally equaled high percentages of votes for Kerry (Hartman: Evidence Mounts Nov 06, 2004)

A Data Analysis done by the University of California, Berkeley, which was replete with full data and had also been reviewed by several professors concluded that *No matter how many factors and variables we took into consideration ... the data show with 99.0% (sic. Tested at 99%, actual figure 99.9%) certainty that a county's use of electronic voting is associated with a disproportionate increase in votes for President Bush.* (Wikipedia: 2004 ... Controversy: 1)

The legality of using electronic voting machines comes into question when one considers that one cannot assure that the vote cast is properly tabulated.

On the whole the evidence sums up to draw the conclusion that malfunctioning of electronic computers is the norm and will continue unless drastic action to eradicate this menace is forthcoming. It is interesting to note that in the 2007 elections to the Scottish Parliament in the U.K., where electronic voting machines were used for the first time, over a hundred thousand ballots were found spoilt—in some electorates the number of spoilt ballots exceeded the majority that declared the winner. *The high number of spoilt ballots made a mockery of the electoral process* says Alec Salmond, the leader of the Scottish National Party (Ritchie: May 04, 2007) This too happened in the UK, a country of technological advancement which leads one to the conclusion of fraud and deceit.

As Michael Parenti says after analyzing the increases in the votes cast in favour of Bush and Kelley:

> The official 2004 tallies showed Bush with 62 million votes, about 11.6 million more than he got in 2000. Meanwhile Kerry showed only 8 million more votes than Gore. To have achieved his remarkable 2004 tally, Bush would need to have kept all his 50.4 million from 2000, plus a majority of new voters plus a large share of the very liberal Nader defectors. (Z Net Daily Commentary)

The total evidence at hand supports the statement of Professor Michael Hout : *No matter how many factors and variables we took into consideration the significant correlation in the votes for President Bush and electronic voting cannot be explained. (Wikipedia. 2004. Controversy: P.35)*

4.1

VOTING MACHINE MANUFACTURERS

Voting Machines Violate your right to vote (Landes)

FOR AN UNBIASED ELECTION to proceed it is essential that the manufacturers of voting machines should not be tainted with political partisanship. But this was not to be.

The scene of voting machine manufacture is dominated by a few manufacturers.

Diebold is a major player with Bob Urosevich as President. In 2002, Diebold acquired Global Election Systems.*Urosevich has headed three different voting machine companies under five different corporate names. (Harris: 2003: 14)*

Global Election Systems which was set up in 1991, had acquired the AccuVote System.

ES&S (Election Systems & Software), another manufacturer has as its Vice President Todd Urosevich, the brother of Bob Urosevich, the President of Diebold. ES&S voting machines count around 50% of all votes in the USA. ES&S and Diebold together count and tabulate easily over 80% of all ballots in the USA.

Another manufacturer is Sequoia, a company controlled by De La Rue, a British company.

The Diebold machines were used very widely. By November 2002, 33,000 machines were used in the US. The fact that Diebold supported the Republicans in the US Presidential Election is amply proved by the Chief Executive of Diebold Election Systems, Walden O'Dell writing a fund raising letter in the year before the 2004 election to the effect that he was *committed to helping Ohio deliver its electoral votes to the President. (Gumbel: 2005: 3)*

Diebold makes most of the DRE i.e. direct recording electronic machines, that do not have a paper trail.

The selection and the purchase of Voting Machinery were left in the hands of each County and it became a big money spinner. The Election Assistance Commission struggled hard to have a unified basis but failed. It looks as if the manufacturers wielded their power to be able to influence the elections staff in various counties and States so that they could win orders. The purchases ran into millions of dollars. There are three major vendors and all of them were supporters of the Republican Party. Walden O'Dell the chief executive of Diebold, threw a fund raising party for Vice President Dick Cheyney and raised six thousand dollars. He sent a fund raising letter to over a hundred fellow Republican supporters.

Election voting machine manufacturers have been found responsible for selling malfunctioning machines.

Diebold Election Systems *has agreed to pay an unprecedented $ 2.6 million fine after the State Attorney General found it had lied about equipment sold in California violated State laws. (Ackerman: E-Voting: San Jose Mercury News: Nov 11, 2004)*

Election voting machine manufacturers and their officers have been found to be corrupt. Wikipedia states:

Some managers and or affiliates of each of these (manufacturers) also have criminal records, including cases of computer fraud, embezzlement, and bid rigging. In addition, voting machine companies have been accused of major security and law violations. Employees including senior executives have been found to have had multiple prior convictions including bans for bid rigging, embezzlement, and drug trafficking, installing uncertified and untested versions of software on touchscreen voting machines and Tampering with computer files. (Wikipedia: 2004. Controversy: 6-9)

Diebold retained a company by the name of Scientific Applications International Corporation(SAIC) to assess the veracity of the Diebold software in order to avoid negative publicity that they were getting. Mark Lewellen-Biddle states that in 1993, the Justice Department brought charges against the company for *civil fraud in an F15 Fighter Contract. In May 1995 the company was charged with lying about security system tests it conducted for a Treasury Department currency plant in Fort Worth, Texas. (Lewewllen-Biddle:12/11/2003).* This reveals the responsibility of a company that was entrusted with certifying the security standards of voting machines. This should have been looked into and approval withdrawn.

In the case of Sequoia, it is reported that *its management has a remarkable record of dishonesty; its executives Phil Foster and Pasquale Ricci were convicted in 1999 of paying Louisiana Commissioner of Elections Jerry Fowler an $8 million bribe to buy their voting machines. (Lefebvre: 24/12/2003)*

Arkansas Secretary of State, Bill McCuen pleaded guilty to taking bribes re computerized voting systems (The Baton Rouge Advocate: 02-05-2002)

The Help America Vote Act passed by Congress in order to avoid the problems that beset the 2000 poll in Florida authorized an expenditure of $ 2.6 billion. Whitney even says that *What*

Congress really did was to throw $ 2.65 billion at the States so that they could lavish it on a handful of private companies that are controlled by ultra conservative republicans, foreigners and felons (Whitney, Nov 3, 2004)

What Congress should have done was to insist on the adoption of security systems, the use of secure codes and proper certification before ordering voting machines or allocating funds for their purchase.

Connections between elections staff and the manufacturers of voting machines have been common, and that at a high level of election staffing. There have been undue connections between companies that have certification rights for voting machines and persons running for public office

As already quoted, Walden O'Dell the CEO of Diebold is supposed to have said that he was *committed to help Ohio declare its electoral votes to the President. (Democracy Now: Sept 04, 2003)* Perhaps connected to this comment is Kenneth Blackwell's decision to order Diebold touch screen voting machines, in the face of an earlier decision by the State which was to purchase only optical scan machines that had a paper trail while the voting machines from Diebold did not have any paper trail.

The Dallas County election administrator in 1985, Conny McCormack later became a supporter for the Diebold touch screen machines as registrar-recorder of Los Angeles County. (Gumbel: 2005:194)

Deborah Seiler, the head of the California election division later became a Sales Rep. for Diebold (Gumbel:2005:191)

Sandra Morham, the former Secretary of State, after leaving office became a State lobbyist for ES&S (Gumbel: 2005:230)

Bill Jones, the Secretary of State in California who in 2002 ruled that printers was an optional item and that the elections code of

California which specified that all voting machines should carry a paper version of each ballot cast should not be read too strictly, thus being lenient on the Voting manufacturers, later became a hired consultant for Sequoia. (Gumbel: 2005:243)

Chuck Hagel, a previous chairman of ES&S became a Republican candidate for a Senate seat in Nebraska. This victory was quite unexpected. *Senate victory against an incumbent Democratic Governor was the major Republican upset in the November Election (Washington Post 1/13/1997)* His victory was unexpected because the electorate comprised black communities that had earlier never voted Republican. He had as head of ES&S computerized Nebraska's voting system. The voting was almost entirely done on unauditable machines he had just sold the State. (Hartman: Nov 04, 2005)

As already pointed out electronic voting machines have to be manufactured to precision, they have to be secure and tamper proof and the evidence at hand is to the effect that the manufacturers have totally failed to be above board in attending to the noble task of manufacturing foolproof voting machinery. As Lynn Landes observed:

When voting machines are used, critical parts of the Voting Rights Act can't be enforced under Section 8i of the Voting Rights Act 42, US Code #1973f, Federal Observers are authorized to observe "whether persons who are entitled to vote are being permitted to vote ... and whether votes cast by persons entitled to vote are being properly tabulated. (Landes: Faking Democracy: 4/6/04)

The many instances where voting machines have malfunctioned in the mandatory task of tabulating the votes cast, can be construed to have caused the contravention of the Help America Vote Act. The penalty for violating the provisions of the Act is a fine of $ 10,000 or imprisonment up to 5 years or both, a penalty that could be enforced on the manufacturers of voting equipment for their negligence in manufacturing foolproof voting machinery.

5

THE CERTIFICATION OF VOTING MACHINES

O N PAPER THE CERTIFICATION system seems sound: *The national testing effort is overseen by NASED's Voting Systems Board which is composed of elections officials and independent technical advisors. NASED has established a process for vendors to submit their equipment to an independent test authority for evaluation against the standards. (Wikipedia: 2004. Controversy: 13)*

It is evident that the system failed perhaps due to the high incidence of malfunctioning that required immediate attention.

However there are doubts about the validity of the security system. Dr Rebecca Mercuri, the Assistant Professor of Computer Science at Bryn Mawr College, a leading expert on electronic voting technology has said that:

> No electronic voting system has been certified even to the lowest level of the US Government or international computer security standards such as the ISO Common criteria nor are they required to comply with such standards. Thus no current electronic voting system is secure by the US Government's own standards (Wikipedia: 2004. Controversy: 25)

Importance has to be given to the certification of electronic voting machines. As Professor Mercuri says:

Any programmer can write codes that display one thing on the screen, records something else and prints yet another result. (Mercuri WebSite)

A great deal of thought has been placed on the standards for certification as shown in Session Law 2005-323,. Senate Bill 223 of the General Assembly of North Carolina , Session 2005:

Prior to certifying a voting system the State Board of Elections shall review or designate an independent expert to review all source code made available by the vendor pursuant to this section and certify only those voting systems compliant with State and Federal Law. At a minimum, the State Board's review shall include a review of security, application vulnerability, application code, wireless security, security policy and processes, security/privacy program management, technology infrastructure and security controls, security organization and governance and operational effectiveness as applicable to that voting system.

Regarding the proper functioning of the electronic voting machines the Law states that *the State Board of Elections shall prescribe rules for the Examination and testing of voting systems in a public forum in the county before and after use in an election.*

In certifying a machine it would be necessary for the inspectors to ensure that the source codes are suitable and secure.

It has been pointed out by Fred Grimm that the source codes are not open to public scrutiny. *(Grimm: MiamiHerald: Nov 09, 2004)* In my opinion what is necessary is that the codes should be looked into by the certification inspectors.

William E. Simon & Sons, a company owned by William E. Simon former Secretary of State and his son Bill Simon acquired a controlling interest of Wyle Laboratories, a company that certifies voting machines. In August 2002, William Simon & Sons was convicted of fraud and had to pay $ 78 million as damages. (*Wikipedia: 2004. Controversy: 10*)

Defective electronic voting machines when found have to get repaired. When machines are repaired or new chips are introduced there is no possibility of certification being done prior to use as is evident in many instances. In Volusia County in Ohio, a defective chip was found:

> *A computer chip is getting the blame for some voter problems in Volusia County. Those ballots will have to be re-fed. The defective chip was found Monday morning as poll workers fired up the machine for the last day of early voting.... The chip was escorted by Deputies to Daytona Beach and is in use right now. (The Command Post:Nov 18, 2004:3)*

This type of incident happened in many precincts throughout the country and in all these cases the certification procedures were compromised.

It is even reported that *Diebold workers also reported that the company switched software in Georgia between tests and the 2002 elections. (Lefebvre: Dec 24, 2003)*

It is also important to note that most DRE systems, including ES & S. machines have internal modems that connect them to external computers. It follows that the external computers can control the functions of these machines in any manner they wish. This ridicules the current process of certification.

Another expert, Professor David Dill of Stanford University states:

The ability to install patches or new software that wasn't certified has many risks, including the introduction of new bugs and more opportunities for tampering ... This opens the possibility of customized tampering by people who know exactly which races they want to affect... Of course even if the certified code is frozen it is easy to think of ways that undetectable back doors (for tampering) could be installed in the software so that some one at the election could choose the winner of the election. (Lefebvre: Dec 24, 2003)

It is also important to note that the Report by John Hopkins and Rice Universities states of the Diebold's DRE technology: *it would be simple and inexpensive to buy a similar card and program it to allow a voter to vote as many times as he wanted. Poll workers would have similar opportunities to directly and unverifiably tamper with vote totals. (Lefebvre: Dec 24, 2003)*

The failure to encrypt the transmission between the voting machines and the central computer enables hacking.

The California State Voting Systems and Procedures Panel in April 2004 held Diebold responsible for the sale of voting machines that did not have certification. This included *sale of machines to the four counties without federal and state certification, last minute soft-ware fixes before the March election and installation of uncertified software in voting machines in 17 counties . (Wikipedia: 2004. Controversy: 42)*

The Panel recommended that the Attorney General should consider civil and criminal charges against the company. (Ibid)

It is clear that the entire certification system was placed in the hands of the vendors of the electronic voting machines which practice can be said to be totally irregular. Peckarsky says:

So called independent testing authorities are paid by the electronic voting machine vendors to inspect and certify the software sold by the electronic voting machine vendors. Such software certification is required by most States and local authorities before they will authorize the use of electronic voting equipment. (Peckarsky: Nov 6, 2006)

It can be considered totally irregular for there to be any connection even in a miniscule manner between the manufacturers or the vendors of the voting machines and their certification. The certification should be done by experts approved by the Elections Departments of the State Administration and they should be paid by the Elections Department. The appointment as well as payments made to the certifying institutions being made by the vendors of the voting equipment makes a total mockery of the certification system.

Instances where uncertified voting machines have been used are many. A few are quoted below:

In Sonomish County it is reported that uncertified voting machines were knowingly used. (Crew: Dec. 03, 2004)

In the November 2002 general election for Commissioner, in Scurry County, Texas a victory predicted by the machines for two Republican Commissioner candidates was overturned after a new computer chip was used to tally the votes. *(Houston Chronicle: Nov 08, 2002)* The introduction of a new chip has to be questioned as they had not been certified.

In Comal County, Texas, in 2002, three Republican candidates won by exactly 18,181 votes each. Re this incident, *a former worker in Diebold's Georgia warehouse says the company installed patches on its machine before the State's 2002 gubernatorial election that were never certified by independent testing authorities or cleared with Georgia election officials. (Wikipedia:2004. Controversy: 24)*

In November 2000, in counting the ballots for Precinct 20, at Glenwood Springs, Colorado, it was found that the ES&S software did not function and a corrected chip had to be obtained to tabulate the results. (City of Glenwood: 2000) What is important in this instance is the fact that the new chip that was introduced had not gone through any checking for certification.

In the case of the 2000 US Presidential Election in the Allamakee County, Iowa, when 300 ballots were fed into the machine, it recorded four million. This is reported to have been corrected by ES&S substituting replacement equipment that had not been certified. (*Wall Street Journal*: Nov 17, 2000)

In the Boone County in Indiana, in the November 2003 election, the MicroVote Machines tallied a total of 144,000 votes when only 5352 votes were cast. (Indianapolis Star: November 9, 2003) It is reported that the Boone County Technology Director and a few Micro-Vote techs 'fixed the problem'. The question at issue is that essentially the coding had to be corrected and this was done outside the certification process.

In the November 2002 General Election it is recorded that a new computer chip had to be flown in to rectify an error. (*Houston Chronicle: Nov 8, 2002*) This newly introduced computer chip had not gone through the certifying process.

In Gaston County, when an alarming 12,000 votes were lost without any trace, *the Election Director hired a voting machine technician to upload the county vote totals and did not oversee the process. (North Carolina's Ballot Blues, 12-04-04)* It appears that certification was overlooked.

In South Carolina, on election day malfunctioning voting machines were reported in a handful of precincts in two counties. The authorities were *forced to switch to paper ballots while technicians got the Votronic touch screens from Electronic Systems & Software up and running within 90 minutes. (USA Today: Nov 03, 2004)* What is also important is how certification standards were ensured after the technical adjustment of the software.

Kohno et. al. state that when the State of Maryland got SAIC and Raba, two computer companies to perform independent analyses of Diebold's AccuVote System it was found that the smartcards had been modified *so that a voter could vote more than once. (Kohno et al: 2004:5)* This action reflects the inadequacy of certification standards.

LAWeekly reported that *Diebold has conceded that it violated California Regulations by using uncertified software in the March 2, Election. (http://www.laweekly.com/ink/04/22/ news-exclusive.php)*

In April 2004, Kevin Shelley is reported to have de-certified e voting machines used in 14 counties. He is later said to have re-certified the machines in 10 counties *after election officials promised to put 23 safe guards in place for the presidential election. (San Jose Mercury News: Nov 03, 2004)*

This action of the Secretary of State could be held to be irregular because the re-certification of any voting machine has to be done after a stringent check is done on a certification process by a specialist. This re-certification pending action to rectify the malfunctioning also indicates the fact that the certification process has been very leniently administered. Such leniency could be the fundamental cause for the widespread malfunctioning of voting machines that has surfaced in this study. It is also important to note that Kevin Shelley, the Secretary of State at California is noted as a strict and efficient official while in other States the Secretaries of State have actually been negligent in their duties.

Specifying the software on any electronic machine is something that can be done. In the case of gaming machines in the State of Nevada, *the State has access to all software on file: illegal to use software not held on file by the State ... state inspectors make random unannounced audit visits to compare computer software with those on file.* For certification the State wants public authority certification as compared to voting machines where the voting machine manufacturers select the certifying companies. (Wikipedia: 2004. Voting Machines)

SEALS ON CERTIFIED VOTING MACHINES

ONCE AN ELECTRONIC VOTING machine has been tested and found suitable, it has to carry a seal that is authentic, cannot be duplicated and this seal should be taken care of. This has to be ensured on a mandatory basis and it is up to the State Elections Department to ensure that the seals are intact at all times. While the officials in charge of the polling precincts have to be vigilant that the seals are intact it is the duty of the County Elections Supervisor to ensure that the seals are intact.

In Warren County, Ohio on 11-5-2004 Associated Press reported that the seals were not intact in 22 voting machines-the seals were missing or broken. (*The Command Post: Nov 15, 2004:5*)

An instance of seals being found not intact is reported from evidence in a New York Court:

> *In New York, voting machines surfaced in a contested State Senate race. Elections officials disclosed in Court that seals were missing or broken on 22 impounded voting machines. (AP: Nov 05, 2004)*

Lynn Landes tells of an interesting instance where the Associated Press newscasters have been given undue access to voting equipment. She states that Christopher Bollyn on Election night in Cook County, *spotted an Associated Press employee connecting her laptop to an ES&S computer at the Cook County (IL) Election headquarters. But was she downloading or uploading data. In an interview with this reporter, Bollyn said, "When I asked the AP reporter if she had direct access to the mainframe computer that was tallying the votes, she said yes and then Burnham (a Cook County Official) stepped in and re asked my question for me. Again the answer was 'yes'".* Cook County officials when questioned had said that it was only downloading and not connected to the mainframe. Bollyn disagrees. *(Landes: Did Networks: 1/5/2005)*

This is a definite instance where unauthorized personnel have had undue access to the tabulation of votes. In tabulating, the work has to be done by the certified electronic machines and it will be irregular to allow anyone who is not an election official to have access of any sort to the electronic machines.

Overall the evidence is to the effect that the certification of electronic voting machines has been inefficiently administered and found wanting in many respects which has led to the erosion of democracy.

The legality of using electronic voting machines comes into question when one considers that one cannot assure that the vote cast is properly tabulated and totaled.

It is important to note that this Study has revealed the fact that electronic voting machines have in many instances malfunctioned in the mandatory task of tabulating and totaling the votes cast. Thus as Landes observed the electronic voting machines are guilt of contravening the Help America Vote Act. It is important to note that the penalty for violating the important provision of proper tabulation amounts to a fine up to $ 10,000 or imprisonment upto 5 years or both. It is quite clear that in many an instance the electronic voting machine manufacturers can be held for the malfunctioning of their equipment.

6

CONCLUSION

*There is no hard evidence of who really wins any election
in America. Our voting system has been privatized and
outsourced. Public control oversight is over. At this point in
time America truly fits the definition of a facist State.
(Lynn:www.ecotalk.org)*

Professor Chalmers Johnson states that corrupt election
laws have made it (Congress) into a forum for special interests.
(Johnson:2004)

Goldstein in his book, *Guide to the 2000 Presidential Election*
states that *the majority of Americans now consider Presidential
elections not important enough to participate. (Goldstein: 2000:xi)*
This could perhaps be due to the Electoral College which actually
usurps the democratic rights of the citizens to elect their President.
To add, I would tend to think that instances like the decision of
the 2000 US Presidential Election, where the winner was declared
by a Judicial Process that decided to stop the counting of votes
cast leaves a scar on the democratic process that can never be
erased.

In the 2000 Presidential Election, I was in Fremont, a city
30 miles south of San Francisco and due to my interest in politics
traveled through many areas in Fremont, Palo Alto where people

gathered and made observations at voting precincts. Everything was normal. There was little heightened activity. Before the election, I visited friends and contacts, some of whom had already voted by post. There was little enthusiasm in them as to who would win.

In the 2004 Presidential Election I was touring the northern states of the USA in my own Roadtrek, a camper motorhome. In the days before the date of the election there was no political enthusiasm visible in any form among the people.

On election day, I passed through many precincts where voting was in progress and did not see any heightened political activity. My observation was that the majority of people were totally disinterested in the election. They were aware that the election was in progress and they wanted to vote and that was it. They were disillusioned about the system. The people whom I met in the 2000 Election were more enthusiastic than those I met in the 2004 Election. When I recollected that I had met a particular friend or contact in my 2000 Election travels and commenced chatting to them on American politics, I always drew a blank– they bluntly stated that the 2000 Election was stolen by not counting the votes and they were thoroughly disillusioned with what could happen in 2004—"what will be will be- it is true that we voted, but our vote does not count" they would say.

This situation is inimical for the future of democracy in America. As Robert F. Kennedy Jr. said: *The first thing we have to do is to restore American Democracy. (The Brad Blog: July2006)*

The USA, is the home of democracy– the country where in the modern world, democracy has prevailed for the longest time, despite small interludes where the voice of the people has been obliterated.

To start with the problem of electronic voting machines should be solved. Either the machines should be fail safe and should not malfunction or they should be totally discarded. It is heartening to note that action is bring taken to ensure that there is a Paper Trail in all voting machines.

In February 2005 Senators Barbara Boxer, Hillary Clinton and Representative Stephanie T. Jones introduced the Count Every Vote Act of 2005. This mandates the requirement of a verified paper ballot for every ballot cast in electronic voting machines.

It is only the insistence of a Paper Trail that can ensure the statement of Thomas Jefferson:

They (people) are the safest depository of the ultimate powers of Government

In view of the necessity to have a paper trail in voting machines, to ensure the true elector in polling, Congressman Rush Holt introduced a bill into Congress requiring a voter verified paper ballot be produced by all electronic voting machines. This was co-sponsored by a majority of the members of the House of Representatives. As Thom Hartmann says:

The two-year battle fought by Dennis Hastert and Tom DeLay to keep it from coming to a vote thus insuring that there will be no possible audit of the votes of about a third of the 2004 electorate has fuelled the flames of conspiracy; theorists convinced Republican ideologues now known to be willing to lie in television advertising-would extend their "ends justify the means" morality to stealing the vote "for the better good of the country" they think single party Republican rule will bring. (Hartmann: Nov 04, 2004)

This can be said to be an alarming instance of political sabotage.

It is important to note that the Governor of Florida, Charlie Crist states that Florida will replace all electronic voting machines that do not have a paper trail. (*The Raw Story, Feb 1, 2007*) Even if a paper trail is available the experience of the 2000 US Presidential Election indicates in no uncertain terms that a recount will not always be allowed. The very high incidence of malfunctioning with no possibility of control over the manufacture of electronic voting machines tells me that the only method of finding the true winner in an election is a hand count. It may take a week for the completion of the count but it is a result that can be guaranteed. In my own experience too a manual count is very reliable. It is also less costly considering the millions spent on electronic voting machines and having to purchase updated versions of machinery again and again when there is new technology coming in. A good alternative is like in India, for the electronic voting machines to be manufactured by the Government as a public venture and not allow Democracy to be at the mercy of electronic voting machine manufacturers.

It is my contention that the skepticism engrained in the people of the US that their votes do not matter is due to a number of factors.

Firstly, it is due to the Electoral College which has eaten into the democratic rights of the people. According to the US Constitution it is the Electoral College that decides and there is no guideline that the Electoral College has to follow the wishes of the majority of people and how they have voted. In the 2000 US Presidential Election when the results in Florida were hotting up, the Florida Legislature even decided to elect a new slate of electors, which *disenfranchised the voters who had already voted and elected electors. (Karunaratne: 2004:105).* As pointed out by me *the last ditch stand of the Florida Legislature to vote in a new set of electors was a desperate attempt of the Republican controlled legislature to ensure that the slate of electors would cast their vote for George Bush whatever the popular vote resulted. (Karunaratne: 2004:106).* There should be no room for undue manipulation in any election for public office.

In a few limited instances–in 1876, 1888, 1960, and in 2000 the Electoral College voted in the person that had not got the majority of the popular vote into the presidency.

The Electoral College is totally undemocratic and as opined by Editorial of The Chicago Tribune,

> *Abolishing the Electoral College would be a welcome step toward a more democratic democracy. (Nov 07, 2004)*

It is important to note that direct popular election is the norm in all other American elections. The decision to have an electoral college was based on the thought that the founding fathers were of the opinion that voters would know little about the candidates. This may have been the situation two centuries ago but is not valid today with the development that has taken place in communication, the use of television and newspapers in disseminating information to the public.

One can hark back to the wisdom of John Adams:

> *The people alone have an incontestable and indefeasible right to institute Government. (Article viii, Massachussets Constitution)*

In addition to the erosion of democracy by the machinations of the Electoral College System, lies that very callous and unjustifiable manner in which the 2000 US Presidential Election was stolen by the Judiciary. As Dershowitz states *it was a dignified, but undemocratic resolution behind closed doors by unelected and politically unaccountable judges who are not supposed to be involved in making political decisions (Dershowitz: 2001:91)* There should never be an instance where a manual recount is stopped and not authorized.

To add insult to injury is the manner in which the 2004 US Presidential Election was hijacked not by not counting the votes as happened in the case of the 2000 US Presidential Election, but by the electronic manipulation of the tabulation and totaling of ballots– the content of this book. In the latter case it is very unfortunate for democracy in America that none of the intelligent stalwarts in the Democratic Party could understand and fathom the underhand machinations that were afoot to steal the election electronically. If only they had insisted on a paper trail in every State, then at least a manual recount could have found the real winner. When I stated this to an American friend he retorted that then what happened in the 2000 Presidential Election could happen again.

The two stolen elections–the 2000 and 2004 Presidential Elections have eaten into democracy in America as well as all over in the democratic world and it is up to every American politician, every political activist to voice themselves and to ensure that democracy will triumph.

Political partisanship causes the erosion of the democratic process. It has to be admitted that political partisanship is at the bottom of irregularities in elections in the USA. This is peculiar to the USA.

Instances of political participation are really many. When the chief officers at the helm–the Secretary of State as well as the Commissioner for Elections at the State level and the Supervisors of Elections at the County level are elected officials, elected essentially on a political party basis, political partisanship can never be ruled out. In the 2004 US Presidential Election the Secretaries of State of the States of Florida and Ohio were the Co Chairs of the Republican Party that supported Bush. It can be considered highly irregular for the officer at the helm of the administration in a State to actively participate directly in any political activity.

Kenneth Blackwell was the Co-Chair of the Republican candidate George W. Bush. He also made an unsuccessful bid as the Republican nominee for the Governor of Ohio in the 2006 election. Blackwell even decided that Provisional voting can be done only at the correct precinct when Federal Law entitles anyone to cast one's vote at any precinct. It was only in mid October that a US District Court Judge James G. Carr ruled that such a guideline would violate Federal Law. In his words: 'Blackwell apparently seeks to accomplish the same result in Ohio in 2004 that occurred in Florida in 2000. (http://www.enquirer.com/editions/2004/10/22/loc_blackwell22html) The Federal Law is very clear on this subject and any official be he the Secretary of State should have been found fault with for giving such an order. Blackwell raised as much as $1.09 million for the Bush campaign. The Toledo Blade reported that Blackwell had got over a million $ in campaign contributions from businesses connected with his work. (16/04/2006)

The sumtotal of unauthorized work done for which Blackwell can be held responsible are:

Arbitrary guidelines used to reject registrations,
Unequal distribution of voting machines,
Impossibly tight challenge deadlines,
Diversion of HAVA funding to observers, who were there to challenge voter qualifications,
Failure to conduct recount
Obstruction of judicial review'
Attempt to disbar attorneys who contested the election.
(Freeman, May 19, 2006)

Thus his role as the Secretary of State to conduct an impartial election is thoroughly dubious.

In the 2002 November and December Senate elections in Louisiana the State Commissioner of Elections was a candidate for the US Senate. (Goldstein: 2003:96)

It should happen that if officials who handle the administration of the election wish to participate in politics either themselves or to support a politician then that official should take leave from his or her duties.

Though it would be undemocratic to have restrictions that people in high office should not aspire to stand for election to elected office, there should be provisions to ensure that once elected they are vowed to be non partisan. In Sri Lanka high level officials cannot be members of a political party or participate in politics. However in all ranks of the Public Service, though political partisanship is banned, there are officials with leanings to a political party. What happens in such a case is to sideline that official and not to give him work in elections. In one definite instance known to me the Assistant Elections Officer of a particular District had leanings to a political party and when that political party was losing its seats at the manual counts, he lost his head and was heard to make untoward utterances. The Assistant Returning Officers who were handling the count (I was one of them) had to restrain him from participating in the count. His duties were immediately taken over and handled by the Assistant Returning Officers. If this timely action had not been taken that count would have ended up in the Supreme Court of Sri Lanka on an election petition.

It can be held that the democratic process for election of officials has gone too far in the USA and that this is the cause for political partisanship. I would reiterate what I wrote in my earlier book: It is necessary that at all levels, at the county as well as at the state level and even at the judiciary, the fact that once appointments have been made to positions of importance the holders of positions should know and understand the importance of being fair and non-partisan. *(Karunaratne: 2004:115)*

The holders of important positions should be sufficiently intelligent and educated to understand that once they hold important office they serve the entire nation and not a part of it. This is the one method of ensuing that democracy prevails. Democracy stands sacred and this should be an accepted fact.

Ms. Norton in Congress referring to the highly dubious vote in Ohio stated:

We have got a failsafe for almost everything else from bulletproof vests to backups for computers. Let us fix our system this year including with failsafes for voting to save our Democracy (Congressional Record: House: Jan 6, 2005: H120)

It is hoped that Democracy will again be victorious in the USA.

7
BIBLIOGRAPHY & REFERENCES

All Experts, '2004 *United States Presidential Election Controversy: Vote Suppression* (:http://en.allexperts. com/e/2004_unitedstates_presidential_election_controversy).

Ackerman, Elise, '*E-Voting Maker to pay a big fine: Attorney General claims Diebold Lied*' in *San Jose Mercury News*, November 11, 2004.,

AFP Features, '*Angry about the lost votes of 2000 BlackFlorida town has strong turn out*', November 2, 2004.

AP Wire, '*Ohio Elections Officials Convicted: Two OhioElection Workers found guilty of rigging vote recount*' January 24, 2007.

George H. Axinn, Foreword to *The Administrative Bunglingthat Hijacked the 2000 US Presidential Election*, by Garvin Karunaratne, The University Press of America, 2004.

BBC News, '*Florida Ballot Papers go missing*', October 28, 2004.

Becker, Jo & David Finkell, '*Now they're registered; Nowthey're Not*' in *Washington Post*, October 31, 2004.

Black Box Voting, Vote Tampering in the 21 st Century, North Carolina's Ballot Blues News & Observer: 12-04-04.

Bollyn, Christopher, (American Free Press), '*Bush-CheneyStole US Election*' in *Conspiracy Planet*, December 5, 2004.

'Both Campaigns See Causes for Litigation in Florida' by New York Lawyer, in *Demos, A NetWork for Ideas &Action*, Nov 03, 04.

Business Week: 'Does Your Vote Matter?' June 14, 2004.

Carter, Jimmy,' *'Still Seeking a Fair Florida Vote'*, in *Washington Post*, September 27, 2004.

Chappell, Brown, *'Voting Machines remain unsecured, expert warns'*, *EE Times* October 28, 2004.

Chardy, Alfonso and David Kidwell, *'In South Florida precincts only minor discrepancies found'*, *Kansas.Com, The Wichita Eagle*, November 27, 2004.

Chicago Tribune, *'A Better Way to Elect a President'*, Editorial, November 7, 2004.

Chicago Tribune, *'OHIO: Machine Glitch Gave Bush Extra Votes'*, November 7, 2004.

City of Glenwood: 'Minutes of City of Glenwood Springs Special City Council Meeting': November 9, 2000.

'Coalition Against Election Fraud: *'Counting Electoral Votes'* Joint Session of the House and the Senate: *9News* , November 10, 2004.

Command Post, *'2004 US Presidential Election: Irregularities'* 11/18/2004.

Conaughton, Gig, *'County: 'Sequencing' error created election results snafu: Misfire briefly changed Proposition 72 tally on Website'*, in *North County Times* (SanDiego) December 3, 2004 pg.B7.

Crampton, Thomas, *'Global monitors find fault with U.SElection'*, *International Herald Tribune*, 2004-11-02.

Crew, Levon, *'Re When a Recount is not a Recount'*, in *BlackBoxVoting.Org*, December 3, 2004.

Culmers, Jackie, Sarah Lueck, Chad Terhume, Ann Carrns & Amy Schatz;, *'Despite Fears Voters Faced Few Problems: But Very Long Line: Some Polling Stations Saw Mischief and Confusion: In Ohio a Six Hour Wait'* *Wall Street Journal* November 3, 2004.

Cummings, Jeanne, 'How Bush Camp Won Ohio' in Wall Street Journal, November 19, 2004.

Cummings, Jeanne, 'Early Voting has arrived', Wall Street Journal, November 3, 2004.

Democratic National Committee, Democracy at Risk: The 2004 Election in Ohio. June 2005.

Dershowitz, Alan, M, Supreme Injustice: How the HighCourt Hijacked Election 2000, Oxford University Press, 2001.

Des Moines Register, 'Broken Vote Counting Machines, Moveon Draw Complaints in Iowa', in Demos: A NetWork for Ideas and Action: Nov 03, 04.

Drinkard, Jim, 'Long Lines on Election Day enhance appeal of early votes', in USA Today, November 18, 2004.

Farhi, Paul, 'Final: Iowa Goes to Bush', in San Jose Mercury News, November 6, 2004.

Fitrakis, Bob, 'Election 2004' in Thuthoutorg: 2005-01-03.

Fitrakis Bob, 'None dare call it voter suppression or fraud' November 7, 2004, The Free Press, April 3, 2007.

Frankell, David & Joe Becker, 'Now they are registered: Now they are Not', The Washington Post, October 31, 2004.

Freeman Steve & Josh Mittledorf, 'A Corrupted Election: Despite what you may have heard the exit polls were right', February 15, 2005.

Freeman, Steven F. 'The Unexplained Exit Poll Discrepancy' pdf (http://www.buzzflash.com/alerts/04/11).

Freeman Steven F, 'Who Really Won-and Lost-the 2004 US Presidential Election', Paper presented at the 61 st Annual Conference of the American Association for Public Opinion Research, Montreal, Canada, May 19, 2006.

Friedman, Brad, 'A Diebold Insider Speaks', September 15, 2005.

Friedman, Brad, 'Diebold Voting Machine Key Copied from Photo at Company's own Online Store', January 24, 2007.

Friedman, Brad, 'Florida Governor Bush Expresses Concern about State Election Systems in light of Leon County Hack Test'. Blogged by Brad Friedman, December 18, 2005.

Friedman, Brad, 'Riverside County CA Supervisor bets a Thousand to One that Sequoia Voting Machines Can't be hacked', Blogged by Brad Friedman, December 6, 2006.

Fund, John, Stealing Elections: How Voter Fraud Threatens our Democracy Amazon, 2004.

Goldstein, Michael L. Guide to the 2004 Presidential Election, Congressional Quarterly Inc., 2003.

Goldfarb, Zachary A, 'A Single Person Could Sway an election' in Washington Post.Com, June 28, 2006.

Grimm, Fred, Hackers Rigging voting machines a real impossibility, in The Miami Herald.Com, November 9, 2004.

Gross, Doug, 'Rights Group Lists Voting Flaws In 7 Southern States', in Plain Dealer, November 6, 2004.

Gumbel, Andrew, Steal This Vote:Dirty Elections and the Rotten History of Democracy in America, Nation Books NY, 2005.

Harris, Bev, Black Box Voting: Ballot Tampering in the 21 st Century, Plan Nine Publishing, 2003.

Hartman, Thom, 'Evidence mounts that the vote May have been Hacked' in Common Dreams Org. Nov 6, 2004.

Hartman, Thom, 'The Ultimate Felony against Democracy', in Common Dreams.Org, Nov 4, 2004.

Hill, Steven, 10 Steps to Repair American Democracy, Poli Point Press, 2006.

House Judiciary Committee Demands Explanation of Irregularities in Ohio' Dec 4, 04.

Houston Chronicle, 'Florida to Replace Missing Absentee Ballots', Demos A NetWork for Ideas & Action; Oct 28, 04.

Hunter, 'Ohio Provisional Ballots: Recounts & Fraud', Daily Kos, November 5, 2004.

Hursti, Harry, *The Black Box Report, : Blackbox Voting. Org.*

International Herald Tribune 'Ohio Election Workers Convicted of Manipulating 2004 US Presidential Recount' January 24, 2007.

Infernal Press.Com, *'How George W. Bush Won the 2004 PresidentialElection'* (http://infernalpress.com/Columns/election. html)

Johnson, Chalmers, *The Sorrows of Empire*, Metropolitan Books, 2004.

Kallestad, Brent, *'Clemency Board to Hear ClemencyAppeals' in The Ledger: The Ledger.Com*, June 17, 2004.

Karunaratne, Garvin, *The Administrative Bungling that Hijacked the 2000 US Presidential Election*, The University Press of America, 2004.

Kennedy, Robert F, & Farhad Manjoo, *'Was the 2004 Election Stolen?', Salon.Com:* June 6, 2006,.

Knapp, Joe, *Count the Vote Rally?, Columbus , Ohio*, January 01, 2004.

Kohno, Tadayoshi, Adam Stubblefield, Aviel D. Rubin & Rachel Konrad, *'Reports of electronic voting trouble top 1,000', in USA Today*, Nov 4, 2004.

Landes, Lynn, *'Did Networks Fail Exit Polls, While AP accessed 2,995 Mainframe Computers?',* 1/5/2005. *(Lynnlandes@ earthlink. net)*.

Landes, Lynn, *'Two Voting Companies & Two Brothers will count 80% of US Election-Using Both Scanners and Touch Screens', Lynn Investigates*, 4/27/04.

Landes, Lynn, *'Republicans Walk Out of Federal Hearing On Voting Machines—While Some Civil Rights Groups Support 'Paperless' Elections' in Lynn Investigates*, 4-13-2004.

Landes, Lynn, *'Faking Democracy-Americans Don't Vote. Machines Do. & Ballot Printers Can't Fix That' in Lynn Investigates*, 4-6-04.

Laner, Thomas W., 'The Risk of E Voting', Electronic Journal of e-Government, Vol.2, Issue3, 2004.

Lange, Werner 'More Votes than Voters in Ohio: Absentee Vote Inflated, Certified Vote in Doubt', Home Blogs Bob Fertik's Blog. December 12, 2004.

Lefebvre, Alex, 'US Voting Machines: Will 2004 Elections be electronically rigged?' in World Socialist Website, Dec 24, 2003.

Mobile Register, 'Voting Snafu Answers Elusive', January 28, 2003.

Levellen-Biddle, Mark, 'Voting Machines Gone Wild', in INTheseTimes, 12/11/2003.

Levon Crew, 'When is a recount not a recount', in Black Box Voting.org. Dec 3, 04.

Liddle,Elizabeth, 2004 Presidential Florida By County ByVoting Machine Type Election Analysis (http://ustogether.org/election 04/LiddleAnalysis.html.

Liptak, Adam, 'Voting Problems in Ohio Set Off an Alarm' in New York Times, November 7, 2004.

Lonewacko Blog, '2004 Presidential Election: Update: November 11, 2004: Lonewacko Blog Weblog.

Lott, John R. Jr. 'Hacker Hysteria' in The Washington Times (washingtontimes.con20040510).

Lynn Investigates, 'Voting and Politics: It's Dirt Simple. If Americans want every vote to count then Americans, not machines must count every vote' in LynnInvestigates-www.ecotalk.org/VotingSecurity.html

McCarthy, Shawn, 'Could Ohio become the Florida of 2004?' in The Globe & Mail, October 20, 2004.

Mercury, Rebecca, Rebecca Mercury's Statement on Electronic Voting, (http://notablesoftware.com:2001).

Msnbc News, 'Machine Glitch Gave Bush extra Ohio Votes: '*Officials say 3893 Vote Error did not affect State result*', in *The Associated Press*, November 5, 2004.

News Analysis from the Washington Despatch, '*Palm Beach County Logs 88,000 more votes than voters*', TWD Wire, November 5, 2004.

'*New York Times Bush Secured Victory in Florida by Veering From Beaten Path*', November 7, 2004,.

New York Lawyer, in Demos, *A NetWork for Ideas & Action*, Nov 03, 04.

Nolan, John, *Return of the Hanging Chad: Recount Continues in Ohio*, in *Boston Globe*, December 16, 2004.

Parenti, Michael, '*The Stolen Election of 2004*', Z Net Daily Commentaries, July 3, 2006.

Peckarsky, Peter, Ron Baiman & Robert Fitrakis, '*Ohio Election System Added Votes Never Cast: Audit LogMissing*', Scoop Independent News, November 6, 2006.

Pitt, Will, '*Voting Fraud in the 2004 Presidential Election*' (*www.ideamouth.com/voterfraud.htm*).

Ritchie, Ken, '*Inquiry Launched into Spoilt Ballot Papers*' in *Politics.Co.UK*, May 04, 2007.

Rockwell, Teed, '*Ohio Vote Fraud: More Bush "Voters" than Residents*' in *Conspiracy Planet:* December 5, 2004.

San Jose Mercury News, '*Activists Seek New look at Ohio Vote*', November 30, 2004.

San Jose Mercury News, '*Balloting Proceeds Despite Problems*'. November 3, 2004.

San Jose Mercury News, ' *Electronic Voting Meets Few Problems in Region*', November 3, 2004.

Schade, Linda, '*Diebold Admits Vote Software used in Maryland Primaries Did Not Meet Fed Standards*', *http://baltimore chronicle.*

com, 08/03/2004.

Sherman, Deborah, *'Team Investigation Uncovers Voter Registration Fraud'*, 9News, 2004-10-11.

Simon, Jonathan D, *'The 2004 Presidential Election: WhoWon the Popular Vote? An Examination of theComparative Validity of Exit Poll and Vote Count Data'*, Verified Vote 2004.

Solon, Diane, *'Young Voters Take Pride in participation'The Plain Dealer*: November 7, 2004.

State News (Michigan State University) ' Smooth Sailing: Voters who were relieved to avoid Electoral College chaos should keep longer memories', Editorial, November 9, 2004.

The Associated Press, 'E-Voting Passes Big Test, Mostly Trouble Free, Voting Companies Pleased, Computer Scinetists Remain Skeptical' , November 4, 2004.

The Command Post '2004 US Presidential Election: Irregularities' November 18, 2004,www.command post.org 2004/2, 11/27/2004.

The Crisis Papers, www.crisispapers.org.

The News & Observer, 'Winners May be Losers', November 9, 2002.

The Raw Story, 'Florida to give paper receipt in 2008Elections, Governor Says', February 1, 2007.

USA Today, 'E Voting Irregularities Raise eyebrows, blood pressure', Nov 03, 2004.

USA Today, 'Voting System Failures', Editorial, November 4, 2004.

USCountVotes.org, *'Pinpointing Precincts with Vote Counting Errors'. Surprising Florida Presidential Election Results'*, November 24, 2004.

Verified Voting.Org, 'The (design) fix is in'. March 16, 2004 .

Dan S. Wallach, *'Analysis of an Electronic Voting System'*, IEEE Symposium on Security and Privacy, 2004.

Wall Street Journal, 'A Comedy of Errors at the Polls' Wall Street

Journal, November 2, 2004.

Welsh, William, *'E-Voting Officials Gather lessons learned from 2004 Presidential Election: Verified Voting Foundation.Org:* November 22, 2004.

Whitney, Mike, *'Election 2004:"Sour Grapes" or Voter Fraud',Election Fraud-USA Presidential Election 2004* (http://www. growrichwhileyousleep.) November 03, 2004.

Whoriskey, Peter, *'Election Whistle Blower Stymied by Vendors'* in *Washington Post,* March 26, 2006.

Wikipedia, *US Presidential Election 2004.*

Wikipedia, *2004 US Presidential Election Controversy & Irregularities.*

Wikipedia, *2004 US Presidential Election Controversy.*

Wikipedia, *2004US Presidential Election Controversy: Exit Polls.*

Wikipedia, *2004 United State Presidential Election Controversy: Ohio.*

Wikipedia, *US Presidential Election 2004 Controversy: Vote Suppression.*

Wikipedia, *2004 United State Presidential Election Controversy: Voting Machines.*

World Socialist WebSite, Editorial Board, *'Allegations of Vote Fraud in Ohio, Florida: Was the 2004 Presidential Election Stolen?'* WSWWebsite, November 24, 2004.

Kim Zetter, *'E Vote Machines Faces Audit', Wired News* Aug 12, 2003.

8

THE AUTHOR: BIOGRAPHICAL SKETCH

Garvin Karunaratne graduated with honours at the University of Sri Lanka, Peradeniya in 1954 and entered the Sri Lanka Administrative Service, where he worked for 18 years, ending his career at Head of Department level, in charge of a major District. Throughout this 18 years he has handled elections, at first as Presiding Officer of a voting precinct, and later in charge of the total process of election administration, functioning as the District Head, with sole responsibility for the conduct-the holding of elections, counting and declaring the winner.

Moving overseas in 1973, he has lived, studied and worked in four other countries: Britain, Bangladesh, the Bahamas and the United States. He held the position of Commonwealth Fund Advisor to the Ministry of Labor and Manpower in the Government of Bangladesh for the years 1981–1983 and was the Commonwealth Fund Advisor in Youth and Community Education to the Government of the Bahamas in 1976 & 1977. He was Lecturer (Assistant Professor) in Third World Studies and Community Development at The Westminster Adult Education Institute from 1990 to 1995. At Michigan State University he worked as the Research Associate on the US Aid-Michigan State University Non-Formal Education Program in Indonesia.

His Degrees include the B.A. Hons. and the M.A. from the University of Sri Lanka, Peradeniya, the Diploma in Community Development (with Distinction) and the M.Ed from the University of Manchester, U.K., the M.Phil. in Agricultural Economics from the University of Edinburgh, U.K. and the Ph.D. in Non-Formal Education & Agricultural Economics from Michigan State University.

He is the author of a number of books on a variety of subjects-Literature, Public Administration, Community Development, Development Economics, Non-Formal Education and Politics. His most recent book: *How the IMF Ruined Sri Lanka & Alternate Programs of Success* (2006) is in the words of Professor George H. Axinn; *a valuable and timely book that will enable international organizations to arrest the trend of failures.* It is a critique of the Structural Adjustment Program.

His book: *The Administrative Bungling that Hijacked the 2000 U.S. Presidential Election,* The University Press of America (2004) depicts in detail how the elections administration failed to hold a fair election, enabling the Judiciary to step in and usurp the right of the people to vote.

His *Microenterprise Development: A Strategy for Poverty Alleviation and Employment Creation in the Third World: The Way Out of the World Bank & IMF Stranglehold* is a daring critique of the World Bank & the IMF. In the words of Professor Sirimal Vithana of the University of Toronto: *it is a valuable addition to the literature on Development Economics.*

His *Non-Formal Education: Theory & Practice at Comilla* (1984), breaks new ground in education and has been published by an international institute of repute–The Bangladesh Academy for Rural Development at Comilla.

His other publications include: *Administering Rural Development in the Third World,* The University Press, Dhaka (1983) two novels: *The Vidane's Daughter* and *Mukulita Piyumo Ayi Vana Meda Me* (in Sinhala) and The Modern Period of Sinhala Poetry (a critical appreciation).

He has also excelled as a real practitioner in economic development. The Youth Self Employment program of Bangladesh designed and established by him in 1982 today trains and guides annually 160,000 persons till they become commercially viable and is easily the largest employment creation program that had stood the test of time (1982 to 2007) The Program has so far benefited over a million people.

Currently as the Founder Director of the Center for Global Poverty Alleviation and a Consultant on International Development, he is engaged in research and writing.

9
INDEX

www.ingramcontent.com/pod-product-compliance
Lightning Source LLC
Chambersburg PA
CBHW070346300526
45791CB00023B/284